Springer Theses

Recognizing Outstanding Ph.D. Research

Aims and Scope

The series "Springer Theses" brings together a selection of the very best Ph.D. theses from around the world and across the physical sciences. Nominated and endorsed by two recognized specialists, each published volume has been selected for its scientific excellence and the high impact of its contents for the pertinent field of research. For greater accessibility to non-specialists, the published versions include an extended introduction, as well as a foreword by the student's supervisor explaining the special relevance of the work for the field. As a whole, the series will provide a valuable resource both for newcomers to the research fields described, and for other scientists seeking detailed background information on special questions. Finally, it provides an accredited documentation of the valuable contributions made by today's younger generation of scientists.

Theses may be nominated for publication in this series by heads of department at internationally leading universities or institutes and should fulfill all of the following criteria

- They must be written in good English.
- The topic should fall within the confines of Chemistry, Physics, Earth Sciences, Engineering and related interdisciplinary fields such as Materials, Nanoscience, Chemical Engineering, Complex Systems and Biophysics.
- The work reported in the thesis must represent a significant scientific advance.
- If the thesis includes previously published material, permission to reproduce this must be gained from the respective copyright holder (a maximum 30% of the thesis should be a verbatim reproduction from the author's previous publications).
- They must have been examined and passed during the 12 months prior to nomination.
- Each thesis should include a foreword by the supervisor outlining the significance of its content.
- The theses should have a clearly defined structure including an introduction accessible to new PhD students and scientists not expert in the relevant field.

Indexed by zbMATH.

Yuchen Wang

Tsunami Data Assimilation for Early Warning

Doctoral Thesis accepted by
The University of Tokyo, Tokyo, Japan

Springer

Author
Dr. Yuchen Wang
Japan Agency for Marine-Earth Science and Technology
Yokohama, Kanagawa, Japan

Supervisor
Prof. Kenji Satake
Earthquake Research Institute
The University of Tokyo
Tokyo, Japan

ISSN 2190-5053 ISSN 2190-5061 (electronic)
Springer Theses
ISBN 978-981-19-7338-3 ISBN 978-981-19-7339-0 (eBook)
https://doi.org/10.1007/978-981-19-7339-0

This Springer imprint is published by the registered company Springer Nature Singapore Pte Ltd.
The registered company address is: 152 Beach Road, #21-01/04 Gateway East, Singapore 189721, Singapore

Supervisor's Foreword

On March 11, 2011, a giant (M 9.0) earthquake occurred along the Japan Trench off Tohoku and generated a huge tsunami along the Pacific coast of northern Japan. The tsunami caused approximately 18,000 casualties as well as the infamous accident at Fukushima Daiichi nuclear power station. Off the Sanriku coast, Earthquake Research Institute of the University of Tokyo had installed cabled instruments to monitor seismic and tsunami waves. The two bottom pressure gauges recorded the generation and propagation processes of the tsunami before it arrived and devastated the Japanese coast. Such offshore tsunami data could have been used to reduce the tsunami disaster. After 2011, the Japanese government started to construct a high-density network (called S-net) with 150 stations of the seismograph and bottom pressure gauge to detect earthquakes and tsunamis offshore.

In the summer of 2015, Yuchen Wang was a junior undergraduate student at Peking University, and participated in the University of Tokyo Research Internship Program (UTRIP) to spend 6 weeks in my lab. He learned the basics of earthquakes and tsunamis, conducted computer simulations, and participated in fieldwork in the Fukushima prefecture damaged by the 2011 Tohoku tsunami. Through this experience, he chose University of Tokyo as his graduate school, and was selected as the first Global Science Graduate Course (GSGC) student in the Earth and Planetary Science at the Graduate School of Science.

In 2016, Yuchen Wang started his graduate research on tsunami data assimilation. The new offshore dense tsunami instrumentations would make it possible to apply data assimilation, a method popularly used in atmospheric science such as weather forecast, to tsunami detection and forecasting. The first paper of tsunami data assimilation was published in 2015 by Dr. Takuto Maeda, who was then at ERI (now a professor at Hirosaki University). Mr. Wang studied this paper, and modified the computation method by using Green's function. His modification significantly reduces the computational time and makes it possible for real-time early tsunami warning (Chap. 2 of his thesis). He further developed methods to apply the tsunami data assimilation for sparse network (Chap. 3), and to automatically detect the tsunami arrival from offshore wave gauges (Chap. 4). These are important and practical improvements to implement the tsunami data assimilation to real-time tsunami early warning system. He made various tests of his methods using synthetic and real data for various tsunamis around Japan and in the world including North and South Americas, Indian Ocean, and New Zealand. As the most realistic application, he applied the automatic detection and tsunami data assimilation methods to the S-net data from the 2016 earthquake off the Fukushima prefecture (Chap. 5).

With these original and pioneering research presented in this thesis, he was awarded a Ph.D. in September 2021. While most of his research results and the contents of his thesis have been published in many scientific journals, this thesis provides a systematic and comprehensive summary of his research development during his 5 years of graduate work.

Prof. Kenji Satake

Tokyo, Japan
October 2022

Abstract

This book focuses on proposing an early tsunami warning system using data assimilation of offshore data. Tsunami is a natural hazard generated by earthquakes, landslides, volcanic eruptions, etc. A rapid and accurate early tsunami warning is important for disaster mitigation. The tsunami data assimilation approach has been proposed for early tsunami warning. It estimates the tsunami waveform by assimilating observed offshore data into a numerical simulation without considering the source. However, the previous data assimilation approach has a relatively high computational cost as it is necessary to run numerical simulations to obtain the entire tsunami wavefield during the assimilation process. The previous approach also requires a dense offshore observation network, making it difficult to conduct data assimilation in regions with sparse observations.

In this book, I first proposed Green's Function-based Tsunami Data Assimilation (GFTDA) to reduce the computation time for assimilation. It can forecast the waveform at Points of Interest (PoIs) by superposing Green's functions between observational stations and PoIs. GFTDA achieved an equivalently high accuracy of tsunami forecasting as the previous approach while saving sufficient time to achieve an early warning. Next, I proposed a modified tsunami data assimilation method for regions with a sparse observation network. The method uses interpolated waveforms at virtual stations to construct the complete wavefront for tsunami propagation. Its application to the 2009 Dusky Sound, New Zealand earthquake, and the 2015 Illapel earthquake revealed that adopting virtual stations greatly improved the tsunami forecast accuracy for regions without a dense observation network. Additionally, I proposed a real-time tsunami detection algorithm using Ensemble Empirical Mode Decomposition (EEMD). The tsunami signals of the offshore bottom pressure gauge can be automatically separated from the tidal components, seismic waves, and background noise. The algorithm could detect tsunami arrival with a short detection delay and accurately characterize the tsunami amplitude. Furthermore, I combined the tsunami data assimilation approach with the real-time tsunami detection algorithm and applied it to the tsunami of the 2016 Fukushima earthquake. The proposed tsunami data assimilation approach can be put into practice with the help of the real-time tsunami detection algorithm.

Parts of this thesis have been published in the following journal articles:

[1] Wang, Y., Satake, K., Maeda, T., & Gusman, A. R. (2017). Green's Function-based Tsunami Data Assimilation (GFTDA): A fast data assimilation approach toward tsunami early warning. *Geophysical Research Letters*, **44**, 10282–10289. https://doi.org/10.1002/2017GL075307.

[2] Wang, Y., Satake, K., Maeda, T., & Gusman, A. R. (2018). Data assimilation with dispersive tsunami model: a test for the Nankai Trough. *Earth, Planets and Space*, **70**, 131. https://doi.org/10.1186/s40623-018-0905-6.

[3] Wang, Y., Maeda, T., Satake, K., Heidarzadeh, M., Su, H., Sheehan, A. F., & Gusman, A. R. (2019). Tsunami data assimilation without a dense observation network. *Geophysical Research Letters*, **46**, 2045–2053. https://doi.org/10.1029/2018GL080930.

[4] Wang, Y., Satake, K., Cienfuegos, R., Quiroz, M., & Navarrete, P. (2019). Far-field tsunami data assimilation for the 2015 Illapel earthquake. *Geophysical Journal International*, **209**, 514–521. https://doi.org/10.1093/gji/ggz309.

[5] Wang, Y., Satake, K., Sandabata, O., Maeda, T., & Su, H. (2019). Tsunami data assimilation of cabled ocean bottom pressure records for the 2015 Torishima volcanic tsunami earthquake. *Journal of Geophysical Research: Solid Earth*, **124**, 10,413–10,422. https://doi.org/10.1029/2019JB018056.

[6] Wang, Y., Satake, K., Maeda, T., Shinohara, M., & Sakai, S. (2020). A method of real-time tsunami detection using Ensemble Empirical Mode Decomposition. *Seismological Research Letters*, **91**, 2851–2861. https://doi.org/10.1785/0220200115.

[7] Wang, Y., & Satake, K. (2021). Real-time tsunami data assimilation of S-net pressure gauge records during the 2016 Fukushima earthquake. *Seismological Research Letters*, **92**, 2145–2155. https://doi.org/10.1785/0220200447.

[8] Wang, Y., Tsushima, H., Satake, K., & Navarrete, P. (2021) Review on Recent Progress in Near-Field Tsunami Forecasting Using Offshore Tsunami Measurements: Source Inversion and Data Assimilation. *Pure and Applied Geophysics*, **178**, 5109–5128. https://doi.org/10.1007/s00024-021-02910-z.

Acknowledgements

I thank my supervisor, Prof. Kenji Satake, for his support and supervision during my research, from my UTRIP internship to my Master's and Ph.D. courses. I learned the discipline of a great scientist from him. I thank my secondary supervisors, Prof. Takashi Furumura and Prof. Shingo Watada, for their support and valuable comments. I thank my family members in China for their continuous support during my study.

I thank Prof. Takuto Maeda, Dr. Naotaka Chikasada-Yamamoto, Dr. Hiroaki Tsushima, Dr. Tatsuya Kubota, Dr. Masaki Yamada, Dr. Aditya R. Gusman, Dr. Iyan E. Mulia, Dr. Yifei Wu, Dr. Tung-Cheng Ho, Dr. Osamu Sandanbata, and Dr. Satoshi Kusumoto for their assistance. I thank Prof. Rodrigo Cienfuegos, Prof. Patricio Winckler, Dr. Natalia Zamora, Mr. Marco Quiroz, and Ms. Paula Navarrete from CIGIDEN, Chile, for their help and support. I thank Prof. Mohammad Heidarzadeh, Prof. Anne F. Sheehan, Dr. Zhiyuan Ren, Dr. Peitao Wang, and Mr. Heng-Yi Su for their collaboration.

I appreciate the Earthquake Research Institute for providing funding for my Master's courses at the University of Tokyo (Global Science Graduate Course). I thank the Japan Society for the Promotion of Science for providing funding for my Ph.D. courses (KAKENHI 19J20293).

Contents

Abbreviations

3-D NS	Three-dimensional Navier-Stokes
4-D Var	Four-dimensional variational assimilation
ANN	Artificial Neural Network
CLVD	Compensated Linear Vector Dipole
DART	Deep-ocean Assessment and Reporting of Tsunami
DONET	Dense Oceanfloor Network System for Earthquakes and Tsunamis
DSFO	Deep Sea Floor Observatory
DSP	Dispersive
EEMD	Ensemble Empirical Mode Decomposition
EIC	Earthquake Information Center
EMD	Empirical Mode Decomposition
EnKF	Ensemble Kalman Filter
GEBCO_2014	General Bathymetric Chart of the Ocean released in 2014
GFTDA	Green's Function-based Tsunami Data Assimilation
IMF	Intrinsic Mode Function
IOC	Intergovernmental Oceanographic Commission
JAMSTEC	Japan Agency for Marine-Earth Science and Technology
JMA	Japan Meteorological Agency
LLW	Linear Long-Wave
LM-IMF	Last-Moment IMF
MOANA	Marine Observations of Anisotropy near Aotearoa
NIED	National Research Institute for Earth Science and Disaster Resilience
NOAA	National Oceanic and Atmospheric Administration
OBPG	Offshore Bottom Pressure Gauge
OBS	Ocean Bottom Seismometer
PoI	Point of Interest
PTWC	Pacific Tsunami Warning Center
S-net	Seafloor Observation Network for Earthquakes and Tsunamis
tFISH	tsunami Forecasting based on Inversion for sea-Surface Height
TRI	Time Reverse Imaging
USGS	United States Geological Survey

Chapter 1
Introduction

1.1 Tsunami Early Warning

Tsunamis are devastating natural disasters that can cause substantial loss of life and severe property damage. The 2011 Tohoku Earthquake (M 9.0) generated a catastrophic tsunami that killed thousands of people and caused serious damage to the nearby Fukushima Daiichi Nuclear Power Station [50]. Tsunami early warning for near-shore communities can help enable timely evacuation. Accurate and rapid forecasting of impending tsunamis is essential for successfully reducing the loss of life and property damage [59].

Generally, there are three types of tsunami forecasting methods for early warning (Fig. 1.1). For most seismogenic tsunamis, the traditional tsunami early warning approach relies on rapid estimates of the earthquake location, depth, and seismic moment [33]. For example, the Japan Meteorological Agency (JMA) issues the first tsunami warning for near-field tsunamis about 3 min after an earthquake [26]. The earthquake location, depth and magnitude are estimated from seismic data, and then used to search tsunami simulation database that stores tsunami calculation results for many earthquake-fault models. The tsunami arrival times and amplitude on 66 forecasting segments of Japanese coasts are used for tsunami warning. In addition, the Pacific Tsunami Warning Center (PTWC) issues tsunami early warnings based on the results of W-phase source inversion [14]. The W phase is a fast, long-period seismic wave used for rapid point source moment tensor inversions [8, 27]. W-phase inversion can be completed within 5–10 min when seismic data at regional distances are used. After an earthquake, the point source moment tensor is obtained using the regional seismic data, and forward simulation is then used to calculate the tsunami wavefield [14, 55, 61]. For the 2015 Illapel earthquake, the PTWC issued a tsunami early warning to the Chilean coast after the earthquake and tsunami warnings or advisories for other regions, such as Easter Island, New Zealand, Fiji, Hawaii, and the Solomon Islands. It later downgraded the warnings. This approach works quickly but is affected by uncertainties in the rapid seismic source inversion [12]. Additionally, some earthquake events have an abnormal focal mechanism (e.g.,

Y. Wang, *Tsunami Data Assimilation for Early Warning*, Springer Theses,
https://doi.org/10.1007/978-981-19-7339-0_1

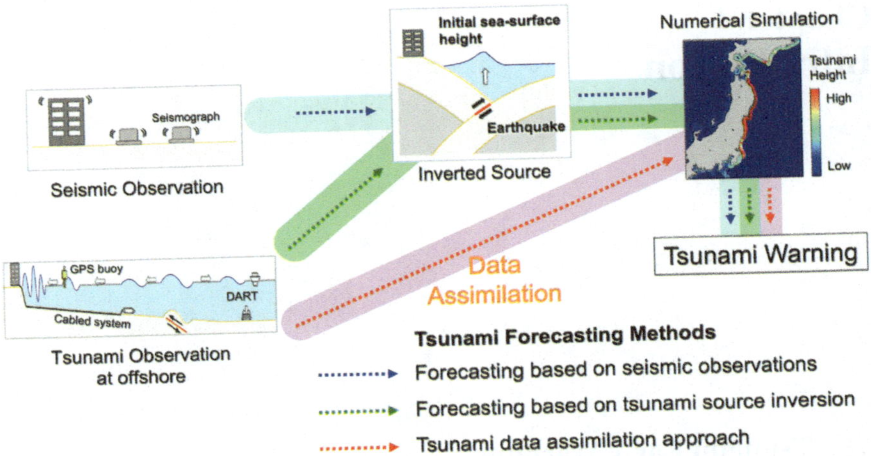

Fig. 1.1 Three types of tsunami forecasting methods. The blue arrow indicates the traditional tsunami early warning method based on seismic observations. The green arrow indicates the method using offshore tsunami data to invert the initial condition. The red arrow indicates the tsunami data assimilation approach. Reprinted from Wang et al. [64] Copyright 2021

Compensated Linear Vector Dipole, CLVD), making it difficult to forecast tsunamis from seismic information [63].

Another tsunami forecasting method uses offshore tsunami data to invert the initial condition, such as the initial slip on fault or initial sea surface displacement. It then simulates the propagation process to forecast the tsunami height or inundation on coasts. For example, in Japan, Tsushima et al. [59] proposed tsunami Forecasting based on Inversion for Sea-surface Height (tFISH). It inverts the tsunami waveform data to estimate the distribution of initial sea-surface height, then derives tsunami waveforms from the estimated source parameters. This method has been successfully applied to retroactive studies of the 1896 Sanriku and 2011 Tohoku earthquakes [59, 60]. In the United States, the National Oceanic and Atmospheric Administration (NOAA) attempted to build practical tsunami forecasting tools based on offshore tsunami data from the Deep-ocean Assessment and Reporting of Tsunami (DART) system, known as the Short-term Inundation Forecast for Tsunamis [57, 58]. Similar methods include the Near-field Tsunami Inundation Forecasting [20], Time Reverse Imaging (TRI) [22], and multi-index methods [65]. However, these methods still must estimate the tsunami source so uncertainties in source estimation affect forecasting. Moreover, both types of methods require the origin time of an earthquake [22, 59], and they do not work for tsunami events caused by non-earthquake origins like landslides or volcanoes. For example, the collapse of the Anak Krakatau Volcano in December 2018 generated a tsunami in the Sunda Strait, Indonesia, with a maximum run-up height of 13.5 m [38]. Inversion methods have difficulties with such non-earthquake tsunamis.

Different from these two types of methods, the tsunami data assimilation approach is a new method of tsunami forecasting that completely avoids the complexities and uncertainties in source estimation. Data assimilation was originally used for operational numerical weather forecasts in meteorology and oceanology [24, 35], and was first proposed for tsunami forecasting by Maeda et al. [36]. The tsunami data assimilation approach forecasts the tsunami based on offshore observations without considering the earthquake source parameters or the initial sea surface distribution. The tsunami wavefield, including tsunami height and velocities in two directions, can be estimated based on real-time data from tsunameters [19, 36]. A major characteristic of the tsunami data assimilation approach is that it does not rely on source information, and is thus free from the complexities of source characterization [12, 67]. This method has already been successfully applied to retroactive studies of synthetic tsunamis in the Tohoku region of Japan, which were generated by an earthquake similar to the 2011 Tohoku earthquake [36], and real tsunamis recorded by pressure gauges in the Cascadia subduction zone [19]. More details on the tsunami data assimilation approach are provided in Sect. 1.3.

1.2 Numerical Modeling of Tsunami Propagation

For either type of early warning, it is important to simulate the tsunami propagation by numerical modeling. In this section, I will introduce the basic governing equations of tsunami propagation. Tsunami is a series of oceanic gravity waves as the gravity is the restoring force. The motion of fluid can be described by the following three-dimensional Navier–Stokes (3-D NS) equations, where we take the z axis in the vertical upward and the x and y axes in horizontal space in Cartesian coordinates [45].

$$\frac{\partial V_x}{\partial t} + V_x\frac{\partial V_x}{\partial x} + V_y\frac{\partial V_x}{\partial y} + V_z\frac{\partial V_x}{\partial z} = -\frac{\partial}{\partial x}\left(\frac{p}{\rho}\right) + \upsilon\nabla^2 V_x \tag{1.1}$$

$$\frac{\partial V_y}{\partial t} + V_x\frac{\partial V_y}{\partial x} + V_y\frac{\partial V_y}{\partial y} + V_z\frac{\partial V_y}{\partial z} = -\frac{\partial}{\partial y}\left(\frac{p}{\rho}\right) + \upsilon\nabla^2 V_y \tag{1.2}$$

$$\frac{\partial V_z}{\partial t} + V_x\frac{\partial V_z}{\partial x} + V_y\frac{\partial V_z}{\partial y} + V_z\frac{\partial V_z}{\partial z} == -\frac{\partial}{\partial z}\left(\frac{p}{\rho}\right) + \upsilon\nabla^2 V_z - g \tag{1.3}$$

Here V_x, V_y, and V_z are velocity components in x, y and z directions, respectively, p is the pressure, ρ is the water density, υ is the kinematic viscosity coefficient for seawater, ∇^2 is the total Laplace Operator, and g is the gravitational acceleration. For large scale motion of water, the effects of viscosity are negligible. We also assume that the sea water is impressible. Equations (1.1)–(1.3) can be simplified as:

$$\frac{\partial V_x}{\partial t} + V_x \frac{\partial V_x}{\partial x} + V_y \frac{\partial V_x}{\partial y} + V_z \frac{\partial V_x}{\partial z} = -\frac{1}{\rho}\frac{\partial p}{\partial x} \tag{1.4}$$

$$\frac{\partial V_y}{\partial t} + V_x \frac{\partial V_y}{\partial x} + V_y \frac{\partial V_y}{\partial y} + V_z \frac{\partial V_y}{\partial z} = -\frac{1}{\rho}\frac{\partial p}{\partial y} \tag{1.5}$$

$$\frac{\partial V_z}{\partial t} + V_x \frac{\partial V_z}{\partial x} + V_y \frac{\partial V_z}{\partial y} + V_z \frac{\partial V_z}{\partial z} == -\frac{1}{\rho}\frac{\partial p}{\partial z} - g \tag{1.6}$$

Equations (1.4)–(1.6) are known as the Euler's equations of motion. When the horizontal scale of motion, or the wavelength, is much larger than the water depth, the vertical acceleration of water is negligible compared to gravity. This means that the horizontal motion of water mass is almost uniform from the ocean bottom to the surface. Such a wave is called a shallow-water wave or a long wave [50]. In such case, the hydrostatic relation is assumed that the pressure is proportional to tsunami height. The 3-D equations become a 2-D problem. The advection terms are neglected. For tsunamis, the typical ocean depth is about 4–5 km, and the source scale of large earthquakes is tens to hundreds kilometers. Therefore, tsunami propagation can be approximated by long-wave approximation. If the tsunami height is small compared to the water depth, Eqs. (1.4)–(1.6) become

$$\frac{\partial V_x}{\partial t} + g\frac{\partial h}{\partial x} = 0 \tag{1.7}$$

$$\frac{\partial V_y}{\partial t} + g\frac{\partial h}{\partial y} = 0 \tag{1.8}$$

Here h is the tsunami height at the sea surface. The equation of continuity, or conservation of mass, can be written as

$$\frac{\partial h}{\partial t} = -d(\frac{\partial V_x}{\partial x} + \frac{\partial V_y}{\partial y}) \tag{1.9}$$

Here d is the water depth. Equations (1.7)–(1.9) form the linear long-wave (LLW) tsunami propagation model.

However, the long-wave approximation breaks down when the wavelength of the water height distribution is not much greater than the water depth [44]. In that case, the dispersive characteristics should be taken into consideration. When the dispersive effects are small ($kh \ll 1$; k is the wavenumber), for the horizontal propagation of waves, the 3-D NS equations can be reduced to a 2-D one by integrating the equations over the water depth [41, 43]. The governing equations of the linear dispersive (DSP) model are derived from the 3-D NS equations, the continuity equation and the kinematic boundary condition at the surface [44, 45]. The continuity equation is the same as Eq. (1.9). However, the difference between the linear DSP model and the LLW model lies in the dispersion terms of the equation of motion [49]. In the LLW model, the dispersion terms are neglected, so the right-hand sides of Eqs. (1.7) and

(1.8) are zero. In the linear DSP model, the equation of motion becomes:

$$\frac{\partial V_x}{\partial t} + g\frac{\partial h}{\partial x} = \frac{1}{3}h^2\frac{\partial^2}{\partial x \partial t}\left(\frac{\partial V_x}{\partial x} + \frac{\partial V_y}{\partial y}\right) \tag{1.10}$$

$$\frac{\partial V_y}{\partial t} + g\frac{\partial h}{\partial y} = \frac{1}{3}h^2\frac{\partial^2}{\partial y \partial t}(\frac{\partial V_x}{\partial x} + \frac{\partial V_y}{\partial y}) \tag{1.11}$$

These equations are also known as the linear Boussinesq equations [50]. I acknowledge that the linear Boussinesq equations or the linear DSP model will deviate from the exact dispersive relation when the dispersion is large ($kh > rsim1$), but it is applicable to seismogenic tsunamis that have not so large dispersion. The nonlinearity is assumed to be small and negligible in the linear DSP model. Saito and Fumurura [45] found that the 2-D tsunami simulation can reproduce the results of the 3-D simulation if we take dispersion effect into account. Though we neglect the vertical motion or variation, the 2-D DSP model is almost same as the 3-D model.

There are two types of boundaries considered in the tsunami computation. The first is a land–ocean boundary. In my study, a total reflection of energy on the coast is assumed. However, for the computation of tsunami inundation, a moving land–ocean boundary is needed. Another kind of boundary is the open ocean boundary along the outer area of the computational region. For the radiation conditions, waves are assumed to leave the computational domain and maintain their slopes [1].

As tsunami is a gravity wave, the conservation of energy flux can be described as

$$b_0 d_0^{1/2} h_0^2 = b_1 d_1^{1/2} h_1^2 \tag{1.12}$$

Here b is the distance between the rays. This equation is known as the Green's Law. It is a classic linear theory result and it describes the evolution of the wave height of periodic waves on plane beaches [54]. The water depth and tsunami height are defined as average values over a horizontal scale of at least one wavelength of tsunami. For plane waves, the ratio b_1/b_0 is 1. Hence, the tsunami height is inversely proportional to one-fourth of the power of water depth change. In previous studies, Green's Law has been used to roughly estimate the tsunami height [3, 47, 61].

In my study, I correct the tsunami height at tide gauges affected by coastal amplification (shoaling effects), because the bathymetry data used in tsunami numerical simulation may slightly deviate from the realistic bathymetry around the tide gauge. However, I must comment here that there is a problem with this application of Green's Law. The Green's Law does not include the effects of partial reflection on the continental shelf. Besides, the water depth or tsunami height should be an average value over more than one tsunami wavelength. This requirement may differ from reality. In my study, the tsunami wavelength varies from ~20 to ~100 km. If we want to obtain the tsunami height at coastal tide gauges more accurately, instead of using Green's Law, it is necessary to obtain bathymetry data with higher resolution. In that case, we could use numerical model to calculate the tsunami height strictly, but such calculation requires finer bathymetry data and higher computational cost.

In addition, in the method of virtual station, which will be explained in Sect. 3.1, the tsunami heights at virtual stations are corrected taking into account the depth difference between the observational stations and interpolation point. The seafloor should be smooth enough to satisfy the assumption that the tsunami propagates as a plane wave and to reduce the effects of partial reflection. However, such correction should still be very cautious because Green's Law requires the average values over a horizontal scale of at least one wavelength.

1.3 Tsunami Data Assimilation Approach

Data assimilation was originally used in meteorology and oceanology. It produces the present state of a system by combining observational data along with an estimated state determined by a forecast model [5, 23, 25]. Several algorithms are proposed for data assimilation, including Optimal Interpolation, Four-dimensional variational assimilation (4-D Var), Kalman Filter, etc. In this section, I briefly review the methods of 4-D Var and Kalman Filter. Then I explain the Optimal Interpolation method that is applied in this study.

1.3.1 Four-Dimensional Variational Assimilation and Kalman Filter

4-D Var provides a least-squares fit of a model trajectory to the background and the observations in an assimilation window. The propagation of the background $\boldsymbol{x}^b$ from the beginning of the window t_0 to the end of the window t_n using the forecast model is represented as:

$$\underline{\boldsymbol{x}^b} = \underline{F}\{\boldsymbol{x}^b(t_0)\} \tag{1.13}$$

where $\underline{\boldsymbol{x}^b}$ is the four-dimensional representation over the assimilation window (the underscore representing a time-dependent vector). In tsunami science, the forecast model F corresponds to tsunami propagation model. The increment is defined as

$$\delta\boldsymbol{x}(t_0) = \boldsymbol{x}(t_0) - \boldsymbol{x}^b(t_0) \tag{1.14}$$

which minimizes a cost function J [16, 56]. In order to improve the conditioning of the cost function, a control variable transform is applied [34] so that the increment is defined as a function of $\boldsymbol{v}$:

$$\delta\boldsymbol{x}(t_0) = \boldsymbol{U}\boldsymbol{v} \tag{1.15}$$

where U is designed such that $UU^T = B$ gives the climatological background-error covariance matrix. Here B is homogeneous and U consists of a scaling by the square root of the estimated power spectrum and a spectral transform. The increment is propagated using the propagation matrix $\underline{M}$

$$\delta\underline{x} = \underline{M}\delta x(t_0) \tag{1.16}$$

The cost function obtained is

$$J(v) = \frac{1}{2}v^T v + (\underline{H}\delta\underline{x} - d)^T R^{-1}(\underline{H}\delta\underline{x} - d) \tag{1.17}$$

where R is the observation error covariance matrix, H is the observation matrix that contains only 0 and 1 values. The innovation vector d is defined as:

$$\underline{d} = \underline{y_n} - \underline{H}(\underline{x^b}) \tag{1.18}$$

Kalman Filter is another data assimilation algorithm. It updates an analysis state from a weighted average of a model prediction and observations of the true state [5]. Define the state variable at time step n to be x_n. This corresponds to the true wavefield, which is unknown but can be estimated. At time $n-1$, the best estimate of x_{n-1} is $\hat{x}_{n-1}$. Let F_n represent how the physical process evolves in one time step from $n-1$ to n. Let P_{n-1} be the covariance matrix of $\hat{x}_{n-1}$. The predicted mean of x_n and its covariance matrix are calculated via the following equations, with the superscript − representing the prior:

$$\hat{x}_n^- = F_n \hat{x}_{n-1} \tag{1.19}$$

$$P_n^- = F_n P_{n-1} F_n^T + Q_n \tag{1.20}$$

where Q_n is a covariance matrix of the system noise that follows a normal distribution. The system noise enables the distribution of the state vector x_n to follow observational data. Even though in some situations, the observational data contain phenomena that is difficult to predict, the system noise can still make the simulation follow observations. It can be included in the model of Kalman Filter. However, because it makes the derivation of the adjoint model of 4-D Var difficult, it is hardly included in 4-D Var. To incorporate data, let H be the observation matrix. Then the distribution (assumed to be normal) of the predicted observations from the prior has mean $H\hat{x}_n^-$ and covariance $HP_n^-H^T$. Let the actual observations be y_n (assumed to be normal) and the covariance of observational error be R_n. To optimally blend the prior estimate and the actual observation, the posterior distribution is calculated as follows:

$$\hat{x}_{n-1} = \hat{x}_n^- + K_n(y_n - H\hat{x}_n^-) \tag{1.21}$$

$$P_n = (I - K_n H)P_n^- \tag{1.22}$$

where the Kalman gain is

$$K_n = P_n^- H^T (H P_n^- H^T + R_n)^{-1} \tag{1.23}$$

Equations (1.19)–(1.23) represent the prediction and update steps in Kalman Filter [48].

Ensemble Kalman Filter (EnKF) is a stochastic approximation of Kalman Filter, which starts with an initial ensemble with N members. There is an ensemble of predicted means $\hat{x}_n^- = [\hat{x}_n^{-(1)}, \ldots, \hat{x}_n^{-(N)}]$ instead of a single estimate at time n. P_n^- is directly calculated as $\mathrm{Cov}(\hat{x}_n^-)$ instead of applying Eq. (1.20). There is also no need to calculate P_n. Instead of storing a full covariance matrix, EnKF presents the same error statistics using an ensemble of model states. N perturbations $v_n^{(i)}$ are added to the observations, which are treated as random variables. For ensemble member i $(i = 1, \ldots, N)$,

$$\hat{x}_n^{(i)} = \hat{x}_n^{-(i)} + K_n (y_n + v_n^{(i)} - H_n \hat{x}_n^{-(i)}) \tag{1.24}$$

As the ensemble size $N \to \infty$, EnKF converges to Kalman Filter. Hence, it leads to an interpretation of the EnKF as a purely statistical Monte Carlo method. The ensemble of model states evolves in state space with the mean as the best estimate and the spreading of the ensemble as the error variance [15].

Though 4-D Var and EnKF methods are effective in estimating the present state, they are computationally expensive. For tsunami early warning purpose, it is quite important to adopt a method with low computational cost to save the calculating time.

1.3.2 Optimal Interpolation

The Optimal Interpolation method is usually adopted for tsunami data assimilation because it has a relatively low computational cost [25, 67]. It is simple and therefore has the advantage of high computational speed for tsunami prediction. It reconstructs the tsunami wavefield from data of offshore tsunami observations by minimizing the total error of all observations [13, 24]. In the numerical simulation, the tsunami wavefield at the n-th time step is represented as a column vector $x_n = (h(n\Delta t, x, y), V_x(n\Delta t, x, y), V_y(n\Delta t, x, y))^T$, where h is tsunami height, and V_x and V_y are velocities in the horizontal directions. The total number of grids is L. The state vector x_n is a $3L \times 1$ column vector, where 1 to L components describe the tsunami height, $L + 1$ to $2L$ components describe the velocity in one direction, and $2L + 1$ to $3L$ components describe the velocity in another direction. The offshore pressure gauges directly provide data of tsunami height, but the velocity components of $\boldsymbol{x}_n$ are reconstructed during the assimilation process.

The data assimilation consists of two steps: a propagation step and an assimilation step. The propagation step is expressed as

$$\boldsymbol{x}_n^f = \boldsymbol{F}\boldsymbol{x}_{n-1}^a \tag{1.25}$$

where $\boldsymbol{F}$ is the tsunami propagation matrix ($3L \times 3L$) such that given the tsunami wavefield at time $t = (n-1)\Delta t$, the forecasted tsunami wavefield at the next time step $t = n\Delta t$ can be calculated. The propagation matrix $\boldsymbol{F}$ corresponds to the tsunami propagation model. Previously, the LLW tsunami model was adopted [19, 36]. In the assimilation step,

$$\boldsymbol{x}_n^a = \boldsymbol{x}_n^f + \boldsymbol{W}(\boldsymbol{y}_n - \boldsymbol{H}\boldsymbol{x}_n^f) \tag{1.26}$$

where the observation matrix $\boldsymbol{H}$ ($m \times 3L$) is a sparse linear matrix, which contains only 0 and 1 values. It extracts the tsunami height at m stations from the forecasted wavefield, and the residual is calculated by comparing it with the real-time observed tsunami height $\boldsymbol{y}_n$, a column vector. Consequently, the residual is multiplied by the weight matrix $\boldsymbol{W}$ ($3L \times m$) to bring the assimilated tsunami wavefield closer to the observed wavefield. The weight matrix $\boldsymbol{W}$ contains the assimilation responses from all stations. Each column represents the assimilation response of one individual station. The weight matrix $\boldsymbol{W}$ is a key controlling factor for the quality of assimilation. It is assumed that the observation matrix $\boldsymbol{H}$ is linear and that the weight matrix $\boldsymbol{W}$ is independent of time. The calculation of weight matrix is based on the correlation distance of the simulation and observation error distributions and the relative magnitude of the observation to simulation errors. We can calculate the weight matrix by minimizing the covariance matrix of $< \varepsilon^a \varepsilon^{aT} >=< (\boldsymbol{x}_n^a - \boldsymbol{x}_n^{True})(\boldsymbol{x}_n^a - \boldsymbol{x}_n^{True})^T >$ as a solution of the linear system:

$$\boldsymbol{W}\left(\boldsymbol{R} + \boldsymbol{H}\boldsymbol{P}^f\boldsymbol{H}^T\right) = \boldsymbol{P}^f\boldsymbol{H}^T \tag{1.27}$$

where $\boldsymbol{P}^f =< \varepsilon^f \varepsilon^{fT} >$ and $\boldsymbol{R} =< \varepsilon^O \varepsilon^{OT} >$ are the covariance matrices of the forward numerical simulation and observations, respectively. Here ε^f and ε^O are the error distributions associated with forward numerical simulations and observations, respectively, which are assumed to be described by Gaussian functions, and ε^{fT} and ε^{OT} are their corresponding transposes [24]. It is assumed that the computational errors are spatially homogeneous on numerical grids, and that the observation errors are not correlated with stations as observations are made independently. These assumptions simplify the matrix $\boldsymbol{R}$ into a diagonal matrix whose components are the standard deviations of the observation errors at each station [36]. A simplified form of Eq. (1.27) can be obtained:

$$\sum_{j=1}^{m} w_{gj}\left(\mu_{ij}^{b} + \delta_{ij}\right) = \mu_{gi}^{b} \tag{1.28}$$

where w_{gj} is components of the weight matrix $\boldsymbol{W}$, μ_{ij}^{b} is the correlation function of errors in numerical forecasts between two observational points (or computational grids). A Gaussian-distributed covariance is assumed for the correlation function of errors, with a characteristic distance of 20 km [36, 62].

By iterating between the propagation and assimilation steps, the tsunami wavefield is gradually assimilated, and the forecasted waveforms at any location inside the modeling domain can be obtained. The point where the tsunami forecast is made is defined as the Point of Interest (PoI). During or after the assimilation process, the forecasted tsunami waveform at a PoI can be obtained based on the forward tsunami simulation from Eq. (1.25) using the present assimilated tsunami wavefield as the initial condition.

1.4 Network of Offshore Bottom Pressure Gauges

Offshore Bottom Pressure Gauges (OBPGs) are installed at water depths ranging from hundreds to thousands of meters to detect tsunamis generated by earthquakes or other causes [6, 9, 52]. They measure the ocean bottom pressure and convert water pressure records into sea surface height [53]. Some data are sent to land in real time, either by satellite or cable. In the United States, the DART system provides an accurate measurement of the tsunami height in the Pacific Ocean through satellite transmission. Offshore of Japan and Canada, there are local dense real-time tsunami observation networks composed of OBPGs, and their data are sent through submarine cables [4, 29].

Off the Pacific coast of Japan, there are two main dense offshore observation networks. In northern Japan, the Seafloor Observation Network for Earthquakes and Tsunamis (S-net) is installed around the Japan Trench [28, 66]. It consists of six groups, with up to 150 stations linked by fiber optic cables [2]. The observatory spacing is approximately 30 km in the east–west direction, which is almost perpendicular to the Japan Trench and approximately 50–60 km in the north–south direction parallel to the trench. The S-net stations are equipped with OBPGs and Ocean Bottom Seismometers (OBSs). There are a pair of identical pressure gauge sensors (HP1WP, HP2WP) at each station to ensure redundancy for tsunami detection [28, 31]. The S-net pressure gauges can detect tsunamis with amplitudes of less than 1 cm [32].

In the Nankai region, the Dense Oceanfloor Network System for Earthquakes and Tsunamis (DONET) is a system for real-time continuous observations and monitoring of earthquakes and tsunamis using an observation equipment network installed on the ocean floor [30]. It was established by the Japan Agency for Marine-Earth Science and Technology (JAMSTEC) and is now operated by the National Research Institute for Earth Science and Disaster Resilience (NIED). There are two phases of the DONET project. DONET1 is located in the Tonankai earthquake hypocenter area, including the offshore of Kii Peninsula and Kumano-Nada. DONET2 is responsible for the Nankai earthquake hypocenter area, including offshore of Shiono Cape and Muroto Cape. The entire system consists of 12 nodes and up to 51 observation points. High-precision observation equipment, including OBPGs and OBSs, are deployed at each observation point at intervals of 15–20 km [29]. The OBPGs of DONET can also detect micro tsunamis with amplitudes less than 1 cm [63].

Besides these two dense observation networks, there are also other distributed OBPGs offshore of Japan. There are two stations of the Deep Sea Floor Observatory (DSFO) off Muroto Cape: PG1 and PG2. JAMSTEC installed them for long-term continuous observations of the phenomena of seafloor changes and seafloor environmental changes. They recorded the tsunamis of the 2015 Torishima earthquake [63]. Moreover, Earthquake Research Institute, the University of Tokyo, installed seismic and tsunami observation systems using seafloor optical fibers off the Sanriku region in 1996 [53]. These OBPGs recorded the tsunami events of the 2011 Tohoku earthquake and the 2016 Fukushima earthquake [18, 60]. The installment of OBPGs enables us to detect tsunamis in real-time operation and perform tsunami data assimilation. Hence, the OBPGs are very important hardware for tsunami early warning.

1.5 Real-Time Tsunami Detection

Real-time tsunami detection plays an important role in tsunami early warning. For the tsunami data assimilation approach, forecasts can be updated continuously as more offshore tsunami data are recorded in real time during the assimilation process. Therefore, real-time tsunami detection is a prerequisite of the tsunami data assimilation approach. Additionally, even for the traditional early warning methods based on seismic observations, the detection of offshore tsunamis is still essential as it helps confirm the actual generation and propagation of a tsunami [11].

The hardware aspect for real-time tsunami detection has advanced significantly in recent years because of the wide installment of OBPGs. As for the software, several algorithms have been proposed for real-time tsunami detection. Although the records of OBPGs are less affected by the complex geometries of bays and harbors, they are usually influenced by tide and ocean currents [42, 44, 51]. Seismic waves and background noise also influence tsunami records. In retroactive studies, the tidal components can be easily removed by subtracting a theoretical tide model, and the tsunami signals can be easily extracted by digital filtering [21, 59]. However, traditional digital filters used in such studies are not applicable to real-time operation. The acausal filter that is usually used in retroactive studies requires future data to filter past data, and the causal filter may change the time of the tsunami peak by the group delay of the filter [10]. Therefore, it is necessary to use a real-time data processing algorithm for automatic tsunami detection. A reliable algorithm for OBPGs should accurately identify the tsunami signals and characterize the tsunami amplitude at a low computational cost and short detection delay.

NOAA developed the tsunami detection algorithm for DART [37]. It first estimates the amplitudes of the pressure fluctuations within the tsunami frequency band and then tests these amplitudes against a 3 cm equivalent sea-level fluctuation threshold value. In NOAA's algorithm, the amplitudes are computed by subtracting predicted pressures from the observations, where the predictions are made using a cubic polynomial fitted to bottom pressure observations over the past 3 h [37]. Additionally, Beltrami [6] used an Artificial Neural Network (ANN) to design a tsunami detection algorithm. The ANN algorithm is based on a two adaptive-weight layer network characterized by logistic sigmoid and linear activation functions. The adaptive weights are obtained from the network's supervised learning. It has a better performance than NOAA's algorithm when the tidal range at the location of interest is higher and the background sea noise is lower [6, 7]. Both algorithms are effective in predicting and filtering out tidal components. However, as the prediction of tides depends on the observation average, an actual propagating tsunami will inevitably influence the predictions. This can affect their performance in characterizing the tsunami amplitude [7]. Moreover, pure predictive algorithms usually fail in reducing the high-frequency components of environmental noise [11].

1.6 Objectives

In this thesis, I mainly focus on four aspects. First, I propose Green's Function-based Tsunami Data Assimilation (GFTDA) to improve the assimilation speed. Then, I adopt virtual stations to perform data assimilation for regions without a dense observation network. Moreover, I propose a new real-time tsunami detection algorithm based on Ensemble Empirical Mode Decomposition (EEMD). Finally, I combine the tsunami data assimilation approach with the real-time tsunami detection algorithm and apply this to S-net pressure gauge records.

The previous tsunami data assimilation approach has a relatively high computational cost because it is necessary to run numerical simulations to obtain the tsunami wavefield. However, the tsunami wavefield of the entire modeling domain needs to be repeatedly calculated during the assimilation process [19], which requires substantial calculation time. Moreover, because of the high computational cost, the adoption of other more complex but realistic models such as the DSP tsunami model is not practical. It may limit the accuracy of tsunami forecasting in some cases where the tsunami dispersive characteristics are evident [46]. In such cases, the long-wave approximation breaks down, and the previous data assimilation approach using the LLW model may not be able to forecast tsunamis with high accuracy. For early warning, the waveforms at the PoIs should be predicted quickly and accurately. In Chap. 2, I propose GFTDA for solving this problem.

Additionally, the previous tsunami data assimilation approach requires enough offshore observed tsunami data to reconstruct the tsunami wavefield [19, 39]. The neighboring stations of an observation network should not be far from each other. Past studies proposed an empirical distance criterion of 50 km to perform tsunami data assimilation [62, 67]. However, an offshore network usually requires a large investment, and many regions with significant tsunami hazards may only afford a few OBPGs rather than a dense network. For example, the Indian Ocean suffered from the 2004 Sumatra–Andaman earthquake and tsunami [17, 40]. However, unlike the Pacific Ocean, there are only sparse tsunami observations in the Indian Ocean. Only six OBPGs are currently available for tsunami detection in the Bay of Bengal (north Indian Ocean). In Chap. 3, I introduce virtual stations between two real observational stations to complement sparse observations.

To put tsunami data assimilation into practice, offshore tsunami data should be provided as the input in real time. In Chap. 4, I describe a novel method of real-time tsunami detection. Unlike the previous predictive algorithms, the method detects the tsunami arrival based on the EEMD without tidal prediction. The tsunami signals are automatically extracted from the original time series of OBPG records. I apply the method to the records of the cabled OBPGs off the Tohoku coast and examine its performance for three earthquakes with different sizes to confirm the validity of the algorithm: the 2016 Fukushima earthquake (M 7.4), the 2011 Tohoku earthquake (M 9.0), and the 1998 off Sanriku earthquake (M 6.4).

In Chap. 5, I combine the tsunami data assimilation approach with the real-time tsunami detection algorithm. Using the 2016 Fukushima earthquake as an example, I forecast the tsunami waveforms at tide gauges in the Sanriku region by assimilating the S-net pressure gauge records. Then, I compare the forecasted waveforms with the real observations at tide gauges to validate the method. Although this study is conducted retroactively, I use the real-time tsunami detection algorithm based on EEMD to process the raw records as if it is operating in real time. In Chap. 6, I discuss the practical implementation and future improvements of the tsunami early warning system.

References

1. Aida I (1969) Numerical experiments for the tsunami propagation-the 1964 Niigata tsunami and the 1968 Tokachi-oki tsunami. Bull Earthq Res Inst 47:673–700
2. Aoi S, Suzuki W, Chikasada NY, Miyoshi T, Arikawa T, Seki K (2019) Development and utilization of real-time tsunami inundation forecast system using S-net data. J Dis Res 14(2):212–224. https://doi.org/10.20965/jdr.2019.p0212
3. Baba T, Hirata K, Kaneda Y (2004) Tsunami magnitudes determined from ocean-bottom pressure gauge data around Japan. Geophys Res Lett 31:L08303. https://doi.org/10.1029/2003GL019397
4. Barnes C, Best M, Johnson F, Phibbs P, Pirenne B (2008) Transforming the ocean science through cabled observatories. Mar Technol Report 30–36. https://doi.org/10.1109/AERO.2009.4839295
5. Bellsky T, Kostelich EJ, Mahalov A (2014) Kalman filter data assimilation: targeting observations and parameter estimation. Chaos 24:024406. https://doi.org/10.1063/1.4871916
6. Beltrami GM (2008) An ANN algorithm for automatic, real-time tsunami detection in deep-sea level measurements. Ocean Eng 35:572–587. https://doi.org/10.1016/j.oceaneng.2007.11.009

7. Beltrami GM (2011) Automatic, real-time detection and characterization of tsunamis in deep-sea level measurements. Ocean Eng 38:1677–1685. https://doi.org/10.1016/j.oceaneng.2011.07.016
8. Benavente R, Cummins PR, Dettmer J (2016) Rapid automated W-phase slip inversion for the Illapel great earthquake (2015, Mw=8.3). Geophys Res Lett 43:1910–1917. https://doi.org/10.1002/2015GL067418
9. Bernard EN, Meinig C (2011) History and future of deep-ocean tsunami measurements. In: OCEANS'11 MTS/IEEE KONA, Waikoloa, Hawaii, pp 1–7. https://doi.org/10.23919/OCEANS.2011.6106894
10. Boore DM, Akkar S (2003) Effect of causal and acausal filters on elastic and inelastic response spectra. Earthquake Eng Struct Dynam 32:1729–1748. https://doi.org/10.1002/eqe.2009
11. Chierici F, Embriaco D, Pignagnoli L (2017) A new real-time tsunami detection algorithm. J Geophys Res: Oceans 122:636–652. https://doi.org/10.1002/2016JC012170
12. Cienfuegos R, Catalán PA, Urrutia A et al (2018) What can we do to forecast tsunami hazards in the near field given large epistemic uncertainty in rapid seismic source inversion? Geophys Res Lett 45:1–12. https://doi.org/10.1029/2018GL076998
13. Daley R (1991) Atmospheric data analysis. Cambridge University Press, pp 98–149
14. Duputel Z, Rivera L, Kanamori H et al (2011) Real-time W phase inversion during the 2011 off the Pacific coast of Tohoku earthquake. Earth Planets Space 63:5. https://doi.org/10.5047/eps.2011.05.032
15. Evensen G (2003) The Ensemble Kalman Fitler: theoretical formulation and practical implementation. Ocean Dyn 53:343–367. https://doi.org/10.1007/s10236-003-0036-9
16. Fairbairn D, Pring SR, Lorenc AC, Roulstone I (2014) A comparison of 4DVar with ensemble data assimilation methods. Q J R Meteorol Soc 140:281–294. https://doi.org/10.1002/qj.2135
17. Fujii Y, Satake K (2007) Tsunami Source of the 2004 Sumatra-Andaman earthquake inferred from tide gauge and satellite data. Bull Seismol Soc Am 97(1A):S192–S207. https://doi.org/10.1785/0120050613
18. Gusman AR, Satake K, Shinohara M, Sakai S, Tanioka Y (2017) Fault slip distribution of the 2016 Fukushima earthquake estimated from tsunami waveforms. Pure Appl Geophys 174:2925–2943. https://doi.org/10.1007/s00024-017-1590-2
19. Gusman AR, Sheehan AF, Satake K, Heidarzadeh M, Mulia IE, Maeda T (2016) Tsunami data assimilation of Cascadia seafloor pressure gauge records from the 2012 Haida Gwaii earthquake. Geophys Res Lett 43:4189–4196. https://doi.org/10.1002/2016GL068368
20. Gusman AR, Tanioka Y, MacInnes BT, Tsushima H (2014) A methodology for near-field tsunami inundation forecasting: application to the 2011 Tohoku tsunami. J Geophys Res: Solid Earth 119:8186–8206. https://doi.org/10.1002/2014JB010958
21. Heidarzadeh M, Satake K (2013) Waveform and spectral analyses of the 2011 Japan tsunami records on tide gauges and DART stations across the Pacific Ocean. Pure Appl Geophys 170:1275–1293. https://doi.org/10.1007/s00024-012-0558-5
22. Hossen MJ, Cummins PR, Dettmer J, Baba T (2015) Time reverse imaging for far-field tsunami forecasting: 2011 Tohoku earthquake case study. Geophys Res Lett 42:9906–9915. https://doi.org/10.1002/2015GL065868
23. Ito S, Nagao H, Yamanaka A, Tsukada Y, Koyama T, Kano M, Inoue J (2016) Data assimilation for massive autonomous systems based on a second-order adjoint method. Phys Rev E 94(4–1):043307. https://doi.org/10.1103/PhysRevE.94.043307
24. Kalnay E (2003) Atmospheric modeling, data assimilation and predictability. Cambridge University Press, Cambridge. https://doi.org/10.1017/CBO9780511802270
25. Kalnay E, Li H, Miyoshi T, Yang S-C, Ballabrera J (2007) 4-D-Var or ensemble Kalman filter. Tellus A: Dyn Meteorol Oceanogr 59:758–773. https://doi.org/10.1111/j.1600-0870.2007.00261.x
26. Kamigaichi O (2015) Tsunami forecasting and warning. In: Meyers RA (eds) Encyclopedia of complexity and systems science. Springer, Berlin, Heidelberg. https://doi.org/10.1007/978-3-642-27737-5_568-3

27. Kanamori H, Rivera L (2008) Source inversion of W phase: speeding up seismic tsunami warning. Geophys J Int 175(1):222–238. https://doi.org/10.1111/j.1365-246X.2008.03887.x
28. Kanazawa T (2013) Japan Trench earthquake and tsunami monitoring network of cable-linked 150 ocean bottom observatories and its impact to earth disaster science. In: 2013 IEEE international underwater technology symposium (UT), Tokyo, pp 1–5. https://doi.org/10.1109/UT.2013.6519911
29. Kaneda Y (2010) The advanced ocean floor real time monitoring system for mega thrust earthquakes and tsunami—application of DONET and DONET2 data to seismological research and disaster mitigation. In: OCEAN 2010.https://doi.org/10.1109/OCEANS.2010.5664309
30. Kawaguchi K, Kaneda Y, Araki E (2008) The DONET: a real-time seafloor research infrastructure for the precise earthquake and tsunami monitoring. In: OCEANS 2008—MTS/IEEE Kobe techno-Ocean, Kobe, pp 1–4. https://doi.org/10.1109/OCEANSKOBE.2008.4530918
31. Kubota T, Chikasada NY, Tsushima H, Suzuki W, Nakamura T, Kubo H (2020) Tsunami analysis using the S-net pressure gauge records during the Mw 7.0 Off-Fukushima earthquake on 22 November 2016 to reduce the effects of tsunami-irrelevant pressure components. In: JpGU-AGU joint meeting 2020, HDS08–11
32. Kubota T, Saito T, Suzuki W (2020) Millimeter-scale tsunami detected by a wide and dense observation array in the deep ocean: fault modeling of an Mw 6.0 interplate earthquake off Sanriku, NE Japan. Geophys Res Lett 47:e2019GL085842. https://doi.org/10.1029/2019GL085842
33. Lomax A, Michelini A (2011) Tsunami early warning using earthquake rupture duration and P-wave dominant period: the importance of length and depth of faulting. Geophys J Int 185(1):283–291. https://doi.org/10.1111/j.1365-246X.2010.04916.x
34. Lorenc AC (2003) Modelling of error covariances by four-dimensional variational data assimilation. Q J R Meteorol Soc 129:3167–3182. https://doi.org/10.1256/qj.02.131
35. Lynch P (2008) The origins of computer weather prediction and climate modeling. J Comput Phys 227:3431–3444. https://doi.org/10.1016/j.jcp2007.02.034
36. Maeda T, Obara K, Shinohara M, Kanazawa T, Uehira K (2015) Successive estimation of a tsunami wavefield without earthquake source data: a data assimilation approach toward real-time tsunami forecasting. Geophys Res Lett 42:7923–7932. https://doi.org/10.1002/2015GL065588
37. Mofjeld HO (1997) Tsunami detection algorithm, Not published paper. http://nctr.pmel.noaa.gov/tda_documentation.html
38. Muhari A, Heidarzadeh M, Susmoro H et al (2019) The December 2018 Anak Krakatau Volcano tsunami as inferred from post-Tsunami field surveys and spectral analysis. Pure Appl Geophys 176:5219–5233. https://doi.org/10.1007/s00024-019-02358-2
39. Mulia IE, Inazu D, Waseda T, Gusman AR (2017) Preparing for the future Nankai Trough tsunami—a data assimilation and inversion analysis from various observational systems. J Geophys Res: Oceans 122:7924–7937. https://doi.org/10.1002/2017JC012695
40. Nalbant SS, Steacy S, Sieh K, Natawidjaja D, McCloskey J (2005) Earthquake risk on the Sunda trench. Nature 435:756–757. https://doi.org/10.1038/nature435756a
41. Nwogu O (1993) Alternative form of Boussinesq equations for nearshore wave propagation. J Waterw Port Coast Ocean Eng 119:618–638. https://doi.org/10.1061/(ASCE)0733-950X(1993)119:6(618)
42. Rabinovich AB (1997) Spectral analysis of tsunami waves: separation of source and topography effects. J Geophys Res: Oceans 102(C6):12663–12676. https://doi.org/10.1029/97JC00479
43. Roeber V, Cheung KF, Kobayashi MH (2010) Shock-capturing Boussinesq-type model for nearshore wave processes. Coast Eng 57:407–423. https://doi.org/10.1016/j.coastaleng.2009.11.007
44. Saito T (2019) Tsunami generation and propagation. Springer Geophysics, Tokyo, pp 1–16. https://doi.org/10.1007/978-4-431-56850-6
45. Saito T, Furumura T (2009) Three-dimensional simulation of tsunami generation and propagation: application to intraplate events. J Geophys Res: Solid Earth 114:B02307. https://doi.org/10.1029/2007JB005523

46. Saito T, Satake K, Furumura T (2010) Tsunami waveform inversion including dispersive waves: the 2004 earthquake off Kii Peninsula, Japan. J Geophys Res: Solid Earth 115:B06303. https://doi.org/10.1029/2009JB006884
47. Sandanbata O, Watada S, Satake K, Fukao Y, Sugioka H, Ito A, Shiobara H (2018) Ray tracing for dispersive tsunamis and source amplitude estimation based on Green's law: Application to the 2015 volcanic tsunami earthquake near Torishima, south of Japan. Pure Appl Geophys 175(4):1371–1385. https://doi.org/10.1007/s00024-017-1746-0
48. Sasaki K, Yamanaka A, Ito S, Nagao H (2018) Data assimilation for phase-field models based on the ensemble Kalman filter. Comput Mater Sci 141:141–152. https://doi.org/10.1016/j.commatsci.2017.09.025
49. Satake K (1989) Inversion of tsunami waveforms for the estimation of heterogeneous fault motion of large submarine earthquakes: the 1968 Tokachi-oki and 1983 Japan Sea earthquakes. J Geophys Res: Solid Earth 94(B5):5627–5636. https://doi.org/10.1029/JB094iB05p05627
50. Satake K (2015) Treatise on geophysics, vol 4, 2nd edn. Elsevier, Oxford, pp 447–504
51. Sato T, Hasegawa S, Kono A, Shiobara H, Yagi T, Yamada T, Shinohara M, Usui N (2017) Detection of vertical motion during a slow-slip event off the Boso Peninsula, Japan, by ocean bottom pressure gauges. Geophys Res Lett 44:2710–2715. https://doi.org/10.1002/2017GL072838
52. Sheehan AF, Gusman AR, Heidarzadeh M, Satake K (2015) Array observations of the 2012 Haida Gwaii tsunami using Cascadia Initiative absolute and differential seafloor pressure gauges. Seismol Res Lett 86(5):1278–1286. https://doi.org/10.1785/0220150108
53. Shinohara M, Yamada T, Sakai S, Shiobara H, Kanazawa T (2016) Development and installation of new seafloor cabled seismic and tsunami observation system using ICT. In: OCEANS 2016 MTS/IEEE monterey, pp 1–4. https://doi.org/10.1109/OCEANS.2016.7761350.
54. Synolakis CE (1990) Green's law and the evolution of solitary waves. Phys Fluids A 3(3):490–491. https://doi.org/10.1063/1.858107
55. Tang L, Titov VV, Moore C, Wei Y (2016) Real-time assessment of the 16 September 2015 Chile tsunami and implications for near-field forecast. Pure Appl Geophys 173:369–387. https://doi.org/10.1007/s00024-015-1226-3
56. Tarantola A (1984) Linearized inversion of seismic reflection data. Geophys Prospect 32:998–1015. https://doi.org/10.1111/j.1365-2478.1984.tb00751.x
57. Titov VV (2009) Tsunami forecasting. In: Bernard EN, Robinson AR (eds) In the sea, ideas and observations on progress in the study of the sea, vol 15. Harvard University Press, Cambridge, pp 371–400
58. Titov VV, Gonzalez FI, Bernard EN, Eble MC et al (2005) Real-time tsunami forecasting: challenges and solutions. Nat Hazards 35:35–41. https://doi.org/10.1007/s11069-004-2403-3
59. Tsushima H, Hino R, Fujimoto H, Tanioka Y, Imamura F (2009) Near-field tsunami forecasting from cabled ocean bottom pressure data. J Geophys Res: Solid Earth 114:B06309. https://doi.org/10.1029/2008JB005988
60. Tsushima H, Hirata K, Hayashi Y, Tanioka Y, Kimura K et al (2011) Near-field tsunami forecasting using offshore tsunami data from the 2011 off the Pacific coast of Tohoku earthquake. Earth Planets Space 63:821–826. https://doi.org/10.5047/eps.2011.06.052
61. Wang D, Becker NC, Walsh D et al (2012) Real-time forecasting of the April 11, 2012 Sumatra tsunami. Geophys Res Lett 39(19):L19601. https://doi.org/10.1029/2012GL053081
62. Wang Y, Maeda T, Satake K, Heidarzadeh M, Su H, Sheehan AF, Gusman AR (2019) Tsunami data assimilation without a dense observation network. Geophys Res Lett 46(4):2045–2053. https://doi.org/10.1029/2018GL080930
63. Wang Y, Satake K, Sandabata O, Maeda T, Su H (2019) Tsunami data assimilation of cabled ocean bottom pressure records for the 2015 Torishima volcanic tsunami earthquake. J Geophys Res: Solid Earth 124:10413–10422. https://doi.org/10.1029/2019JB018056
64. Wang Y, Tsushima H, Satake K, Navarrete P (2021) Review on recent progress in near-field tsunami forecasting using offshore tsunami measurements: source inversion and data assimilation. Pure Appl Geophys 178:5109–5128. https://doi.org/10.1007/s00024-021-02910-z

65. Yamamoto N, Aoi S, Hirata K, Suzuki W, Kunugi T, Nakamura H (2016) Multi-index method using offshore ocean-bottom pressure data for real-time tsunami forecast. Earth Planets Space 68:128. https://doi.org/10.1186/s40623-016-0500-7
66. Yamamoto N, Hirata K, Aoi S, Suzuki W, Nakamura H, Kunugi T (2016) Rapid estimation of tsunami source centroid location using a dense offshore observation network. Geophys Res Lett 43:4263–4269. https://doi.org/10.1002/2016GL068169
67. Yang Y, Dunham EM, Barnier G, Almquistj M (2019) Tsunami wavefield reconstruction and forecasting using the ensemble Kalman filter. Geophys Res Lett 46:853–860. https://doi.org/10.1029/2018GL080644

Chapter 2
Green's Function-Based Tsunami Data Assimilation (GFTDA)

2.1 Principles of GFTDA

GFTDA is proposed to improve the assimilation speed. Green's functions have been used to waveform inversion for fault estimation [26] and TRI for far-field tsunami forecasting [11, 12, 15]. Here I adopt Green's functions in data assimilation for tsunami forecasting. In the Optimal Interpolation algorithm [18] In this method, the residual between the observed and forecasted tsunami height is calculated. If it is non-zero, the station will have an assimilation response, which can be a near-Gaussian shape if the distance between neighboring stations is larger than the characteristic distance of matrix $\boldsymbol{P}^f$ and $\boldsymbol{R}$. I then define the residual vector as $\boldsymbol{x}_n^r \equiv \boldsymbol{W}(\boldsymbol{y}_n - \boldsymbol{H}\boldsymbol{x}_n^f)$, which represents the residual of each station at time $t = n\Delta t$. The residual vector can be rewritten as a linear combination of the unit column vector $\boldsymbol{e}_i^T$ multiplied by the corresponding residual of tsunami height at the ith station r^i:

$$\boldsymbol{x}_n^r = \boldsymbol{W}\sum_i r_n^i \boldsymbol{e}_i^T \tag{2.1}$$

The assimilation response of the ith station can be mathematically represented as $r_n^i \boldsymbol{W}\boldsymbol{e}_i^T$. The assimilation response will propagate across the calculating region. Hence, it will affect the tsunami height and velocity at other grid points. If we use a linear tsunami propagation model, the assimilation response of different steps and stations can be simply superposed. I define the waveform at the jth grid point that results from the propagation of the ith station's assimilation response as the Green's function $G_{i,j}$. In Fig. 2.1, plots of Green's functions $G_{1,s}$, $G_{2,s}$, and $G_{3,s}$ represent the waveforms at PoIs resulting from the propagation of the assimilation responses of the 1st, 2nd, and 3rd stations, respectively. Therefore, by using GFTDA, it is not necessary to simulate the tsunami propagation of the whole region during the assimilation process. Instead, the waveforms at PoIs are synthesized by Green's functions directly.

Y. Wang, *Tsunami Data Assimilation for Early Warning*, Springer Theses,
https://doi.org/10.1007/978-981-19-7339-0_2

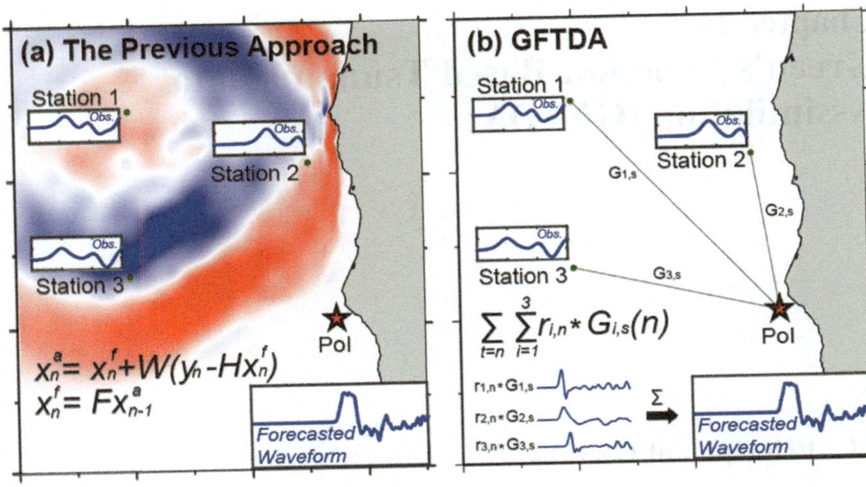

Fig. 2.1 Illustration of the previous data assimilation approach and Green's Function-based Tsunami Data Assimilation (GFTDA). For simplification, I only adopt three offshore observational stations (green dots) to forecast the waveform at the Point of Interest (PoI) (red star). In the previous data assimilation approach, the wavefield of the entire modeling domain is computed. Then, the forecasted waveform at the PoI is recorded in the wavefield. In GFTDA, the Green's functions are computed and stored in advance. Then, the forecasted waveform at the PoI is directly synthesized with Green's functions by matrix manipulation. Both approaches give the same result for the forecasted waveform. Reprinted from Wang et al. [34] Copyright 2017

For a given region with a tsunami observation network (m stations) and N_s PoIs, the number of Green's functions to be calculated is $m * (N_s + m)$. In real practice, the PoIs should include near-shore points because the prediction of tsunami height at these points is most significant to residents. To the contrary, it is not useful to forecast the tsunami at points far away from the coast. Thus, we can omit the computation of Green's functions these points to reduce the required amount of computer memory. In addition, the Green's functions between observational stations themselves are necessary because waveforms of observational stations are updated continuously during the assimilation process.

2.2 Assimilation Process and Mathematical Equivalence

The assimilation process of GFTDA is demonstrated in Fig. 2.2. The Green's functions are calculated by computers or supercomputers in advance. Next, the residual between the observed and forecasted tsunami height is calculated (initial condition is zero) when the tsunami arrives. Then, the residual of each observational station is multiplied with the corresponding Green's functions to forecast the tsunami heights at the observational stations themselves and PoIs for this time step. In the next time

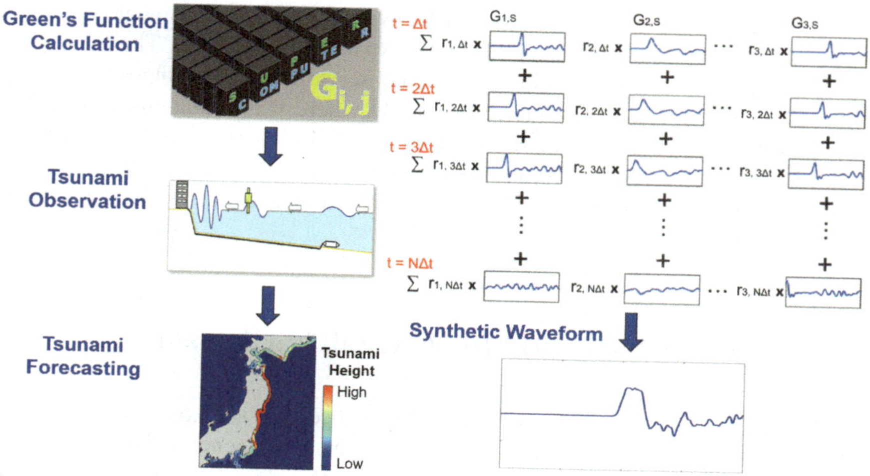

Fig. 2.2 Assimilation process of Green's Function-based Tsunami Data Assimilation. Step 1: The Green's functions are calculated by a computer and stored in advance. Step 2: Offshore stations observe the tsunami height. The residuals between the observed and forecasted height of each station are calculated. The assimilated waveforms at stations and Points of Interests (PoIs) are repeatedly synthesized with Green's functions by matrix manipulation. Step 3: The forecasted waveforms at PoIs are obtained. Reprinted from Wang et al. [37] Copyright 2021

step, the observation data at each station are updated, and the residuals are calculated and multiplied with the Green's functions again. These steps are continuously performed during the assimilation process. The forecast is made using the tsunami data of observational stations and pre-calculated Green's functions. As the tsunami propagates, more observational stations detect the tsunami height and provide more data. Hence, the accuracy of the forecasted waveforms at PoIs gradually increases with time. Using the forecasted waveforms, the tsunami warning level can be decided before the tsunami reaches the shore.

If we use a linear propagation model, GFTDA is mathematically equivalent to the previous tsunami data assimilation approach. The equivalence can be proved as follows.

$$x_n^a = x_n^f + \boldsymbol{W}\left(y_n - \boldsymbol{H}x_n^f\right) = x_n^f + x_n^r \tag{2.2}$$

$$x_n^f + x_n^r = \boldsymbol{F}x_{n-1}^a + x_n^r = \boldsymbol{F}\left(x_{n-1}^f + x_{n-1}^r\right) + x_n^r = \boldsymbol{F}\left(\boldsymbol{F}(x_{n-2}^f + x_{n-2}^r) + x_{n-1}^r\right) + x_n^r = \cdots = \sum_{t=1}^{n} \boldsymbol{F}^{n-t}x_t^r \tag{2.3}$$

$$\sum_{t=1}^{n} \boldsymbol{F}^{n-t}x_t^r = \sum_{t=1}^{n} \boldsymbol{F}^{n-t}\sum_{i=1}^{m} r_t^i \boldsymbol{W}e_i^T = \sum_{i=1}^{m}\left(\sum_{t=1}^{n} r_t^i \boldsymbol{F}^{n-t}\boldsymbol{W}e_i^T\right) \tag{2.4}$$

Equation (2.2) is the assimilation step of previous approach. In a linear system, the propagation of the tsunami wavefield can be expanded as Eq. (2.3). The assimilated waveforms are related to the residuals of each observational station at each time step. The residual vector can be rewritten as Eq. (2.4), where $\boldsymbol{W}e_i^T$ is the unit assimilation response of the ith observational station. Here, $\boldsymbol{F}^{n-t}\boldsymbol{W}e_i^T$ represents the propagation of the assimilation response from the ith station (i.e., Green's function). Therefore, during the assimilation process, we only need to consider the grid points of the stations and PoIs.

2.3 Validation Test—2012 Haida Gwaii Earthquake

On October 28, 2012, an M 7.8 earthquake (M 7.8) occurred at 52.622°N, 132.103°W, off the coast of Haida Gwaii, British Columbia, Canada. It generated a tsunami with a run-up height greater than 3 m. Fortunately, no casualties or major injuries were recorded from the earthquake and tsunami, likely because of the sparse population of the region. The tsunami was recorded by tide gauges and an OBS array with bottom pressure gauges in the Cascadia subduction zone during the Cascadia Initiative Community Experiment [8, 9, 28]. We note that although this network did not transfer data in real time, the recorded data can be adopted to perform tsunami data assimilation retroactively.

2.3.1 *Observed Tsunami Data*

The tsunami data recorded at 26 stations of the Cascadia OBS array [28] are adopted in this study. Gusman et al. [9] applied the previous data assimilation approach to this event, whereas I apply GFTDA to forecast the tsunami retroactively. I applied a band-pass filter with cut-off frequencies of 0.0002 and 0.005 Hz (5000 and 200 s) in order to remove the seismic waves and ocean tides. The pressure data were converted to water height using the hydrostatic relation, and the instrument response was removed from the data. For more details on data processing, please refer to previous studies [8, 9, 28]. I divide the observational stations into two groups: a group for data assimilation of the 15 offshore stations and a group for validation of 11 near-shore stations (PoIs). The locations of these stations are shown in Fig. 2.3a.

2.3.2 *Assimilation Setting*

The *JAGURS* LLW model [3] is used to compute Green's functions. The bathymetry data is derived from the General Bathymetric Chart of the Ocean released in 2014 (GEBCO_2014) [38]. The grid size used in this study in 0.5 arc min. The target area

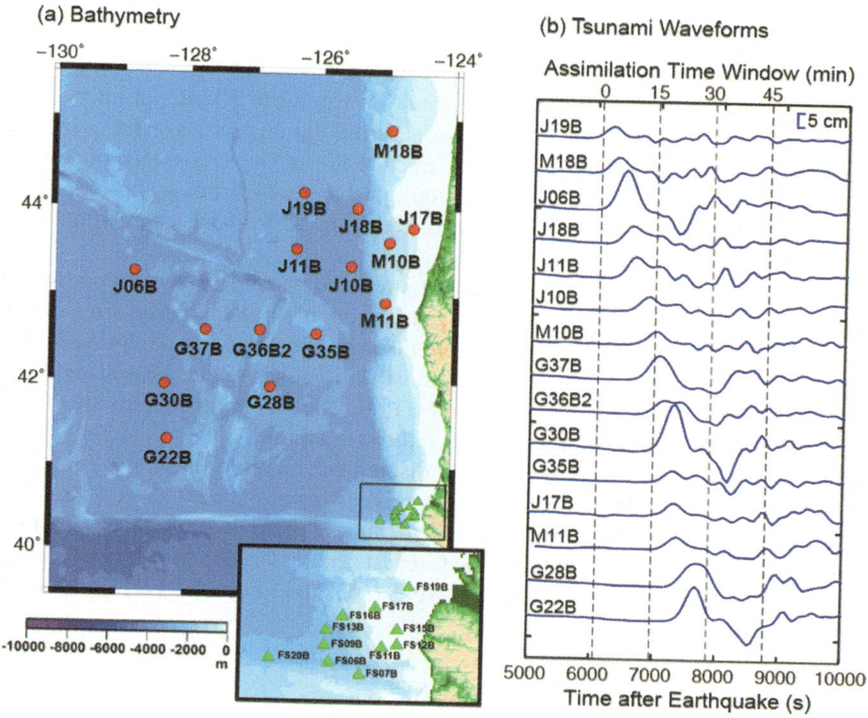

Fig. 2.3 Bathymetry and tsunami waveforms of the observational stations in the Cascadia Initiative Community Experiment. **a** The 15 offshore stations that provide tsunami observation for data assimilation are marked by red circles. Other 11 near-shore stations (i.e., Points of Interest) are indicated with green triangles. I compare the observed waveforms at Points of Interest with the forecasted waveforms calculated by Green's Function-based Tsunami Data Assimilation. **b** The observed waveforms at the 15 offshore stations are plotted on the right panel. Reprinted from Wang et al. [34] Copyright 2017

is 35°N–56°N, 122°W–136°W. The time step of Green's functions is 1 s, consistent with the processed observation data. There are 15 observational stations and 11 PoIs. Therefore, the number of Green's functions computed is $15 \times (11 + 15) = 390$. It does not require a large amount of computer memory.

The earthquake origin time is set as $t = 0.$. The tsunami arrived at the first observational station (J19B) at $t = 6066$ s (Fig. 2.3b). Then, the assimilation process begins. The *assimilation time window* is defined as the period during which the observed data are continuously used for assimilation. That is to say, the waveforms at PoIs are synthesized by GFTDA during the assimilation time window. For example, if the assimilation time window is set to 1800s (30 min), the assimilation process will stop at $t = 7866$ s. The tsunami height at stations is forecasted after $t = 7866$ s.

2.3.3 Results

GFTDA is applied with a series of time windows from 2 to 50 min, at intervals of 2 min. It takes less than 10 s to complete the data assimilation process on the computer system of the Earthquake Information Center (EIC) at the University of Tokyo. Compared to the previous data assimilation approach which takes approximately 1,000 s, GFTDA has a much faster speed, reducing 99% of the calculation time during the assimilation process.

Figure 2.4 compares the observed and assimilated waveforms at 11 PoIs for an assimilation time window of 24 min. In general, the assimilated waveforms match well with observations. At this time, the first tsunami peak has already passed most stations (Fig. 2.3). That is, the data from most stations have been assimilated. The tsunami height is accurately predicted at most PoIs. The maximum heights are underestimated at stations FS06B and FS09B, and the maximum height is slightly overestimated at station FS19B. However, the discrepancy is quite small, so it is fair to say that GFTDA provides accurate tsunami forecasting at the near-shore PoIs.

I calculate two indexes to quantitatively evaluate the performance of GFTDA: forecast accuracy [8] and score [32]. The forecast accuracy is obtained by calculating the geometric mean ratio (K) of the observed (H_o^{arr}) and forecasted (H_f^{arr}) first-peak amplitude [1]:

$$\log(K) = \frac{1}{N_s} \sum \log\left(\frac{H_o^{\mathrm{arr}}}{H_f^{\mathrm{arr}}}\right) \tag{2.5}$$

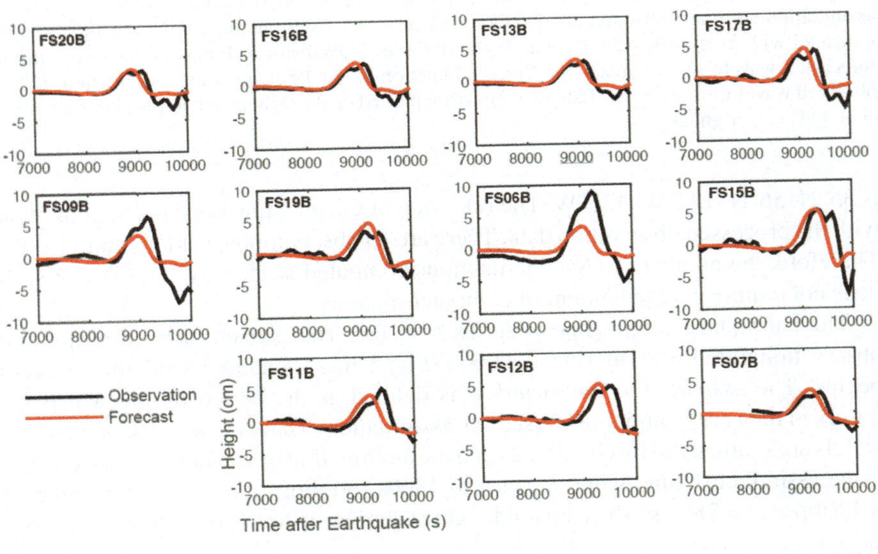

Fig. 2.4 Observed and forecasted waveforms at 11 Points of Interest. The assimilation time window is 24 min. Reprinted from Wang et al. [34] Copyright 2017

$$Accuracy(\%) = \frac{1}{K} \times 100\%(K \geq 1)\ or\ K \times 100\%(K < 1) \quad (2.6)$$

Here N_s is the number of PoIs. The score is calculated as follows:

$$Score(\%) = \left[1 - \frac{\sum (H_f^{arr} - H_o^{arr})^2}{\sum (H_o^{arr})^2}\right] \times 100\% \quad (2.7)$$

Figure 2.5 shows the forecast accuracy and score at different assimilation time windows. In the beginning, both forecast accuracy and score are low. The reason is that the input data is too short to provide accurate forecasting. At this time, the first tsunami peak has not passed many stations. Then, as the time window becomes longer, a sharp increase appears in both the forecast accuracy and score. For a 24 min time window, the forecast accuracy is over 90% and the score is over 80%. After 24 min, the increase in forecast accuracy and score becomes smaller, and there is almost no improvement after 30 min. At this time, as the first tsunami peak has already passed all 15 offshore observational stations, the forecast accuracy and score are saturated.

Hence, the results show that the accuracy of tsunami forecasting is dependent of the assimilation time window. A short time window results in a quick forecast, but the accuracy is low, and the confidence of tsunami early warning remains doubtful. Conversely, if the assimilation time window is very long, although the tsunami forecasting will be more accurate, there may not be sufficient time for those living along the coast to evacuate. Therefore, it is important to select a suitable assimilation time window to strike a balance between forecast accuracy and computation time

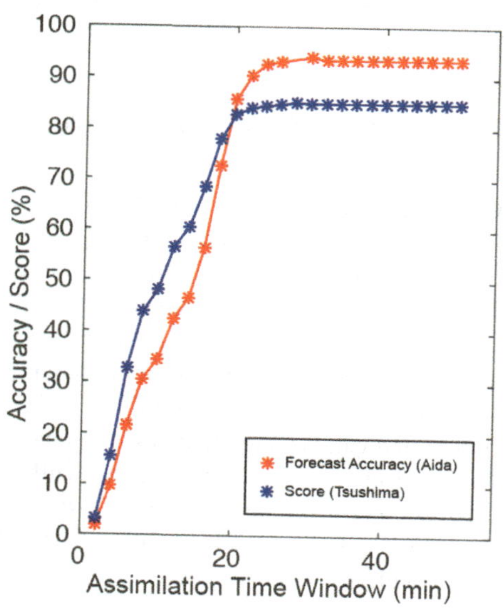

Fig. 2.5 Forecast accuracy and score at different assimilation time windows. The accuracy (Aida's method) and score (Tsushima's method) both show a sharp increase initially and tend to saturate after a 30 min assimilation time window. Reprinted from Wang et al. [34] Copyright 2017

to produce reliable tsunami forecasting. In this study, a time window of 24 min is a proper choice. While it provides a reliable tsunami warning, coastal residents would have approximately 25 min for evacuation. Moreover, it is important to repeat the forecast over time to improve accuracy. Because GFTDA is able to forecast the tsunami instantaneously, it is very effective for repetitive forecast.

For the case study of the 2012 Haida Gwaii tsunami, the score proposed by Tsushima et al. [32] tends to assess the forecasting performance better than the forecast accuracy (Aida's criterion). The sum of the logarithm as in Eq. (2.5) is adopted to calculate the forecast accuracy. If some observations are higher than the forecasted values, and some other observations are lower, the positive and negative logarithm terms will cancel out. Thus, $\log(K)$ can still be very small, resulting in an abnormally high accuracy despite the fact that the residuals between the forecasts and observations are large. In Fig. 2.5, I remove a point (28 min) where the calculated forecast accuracy is abnormally high. Conversely, the score calculation includes the squared error (Eq. 2.7). The errors of different PoIs cannot be canceled out. Here I compare two criteria to evaluate the performance thoroughly. As a general practice of accuracy analysis, it is important to select an appropriate criterion to evaluate the performance fairly.

2.4 Adoption of Linear Dispersive Model—2004 Off the Kii Peninsula Earthquake

This section applies GFTDA to synthetic tsunami data of the 2004 off the Kii Peninsula Earthquake. I calculate Green's functions with both the LLW and linear DSP models and assimilate the tsunami data with both models.

2.4.1 2004 Off the Kii Peninsula Earthquake

Giant earthquakes have occurred along the fault between the Philippine Sea Plate and Eurasian Plate [6, 24], includomg the 1944 Tonankai (M 8.0) and the 1946 Nankai (M 7.9) earthquakes. Additionally, a large (M 7.4) earthquake occurred off the Kii Peninsula in 2004 within the subducting Philippine Sea Plate.

The 2004 off the Kii Peninsula earthquake occurred on September 5, 2004. It was an intraplate event occurring in the outer rise of the Philippine Sea Plate. The main shock at 14:57 UTC followed the foreshock at 10:07 UTC. The tsunami generated by the mainshock showed evident dispersive characteristics [24]. Dispersive waves have a strong directional dependence with respect to the fault strike [24]. The tsunami dispersion is strong in the short-axis direction of the source fault where the tsunami wavelength is short, and it is also strong in the direction along the deep sea where the tsunami travels fast.

GFTDA enables us to use a more complicated tsunami propagation model for data assimilation, as long as the model is linear. I adopt the linear DSP model for data assimilation and compare it with the LLW model.

I compute Green's functions between observational stations and PoIs with both the LLW and linear DSP models. The propagation matrix $\boldsymbol{F}$ is more complicated when using the linear DSP model. Although the computation of Green's functions takes a long time, this step is done in advance and does not affect the efficiency of tsunami data assimilation. During the assimilation process, the forecasted tsunami waveforms can be directly synthesized by multiplying the residual with corresponding Green's functions.

2.4.2 *Synthetic Tsunami Data*

DONET1 was finished in 2011 and DONET2 started its operation in 2015. Hence, real tsunami records of the 2004 off the Kii Peninsula earthquake are not available. In this study, I use synthetic tsunami waveforms to assess the ability of the data assimilation approach.

First, I adopt a source model similar to the mainshock of the 2004 off the Kii Peninsula earthquake for simulation. The fault parameters are determined according to the results of Yamanaka (data available at http://wwweic.eri.u-tokyo.ac.jp/sanchu/Seismo_Note/2004/EIC153.html), which were obtained by analysis of the teleseismic body waves. The epicenter is 137.142°E, 33.143°N. The depth is 10.0 km. The fault direction of the source model is perpendicular to the trough axis with a strike of 135°. The dip and rake angles are 40° and 123°, respectively. The length and width of the rectangular fault are 50.0 and 30.0 km, respectively. The fault slip is 6.5 m, consistent with the magnitude of the mainshock (M 7.4).

Okada's model is applied to calculate the initial sea surface elevation [21], as the initial condition for tsunami numerical simulation. Then, the *JAGURS* model is used to model the tsunami propagation. A total reflection is assumed at the coastline. Sea-bottom friction is not considered in my simulation. To make the tsunami propagation closer to the real situation, the linear DSP model is applied. The synthetic waveforms at observational stations and PoIs are recorded during the numerical simulation.

DONET has 12 nodes, and each node is linked with several observation points. I take one point for each node for Node C, for which I take two points. The intention is to build an evenly-distributed observation network for tsunami data assimilation. Moreover, I use two submarine stations belonging to the DSFO off Muroto Cape: PG1 and PG2. In total, there are 15 observational stations for data assimilation (Fig. 2.6).

Additionally, nine PoIs near population centers on Shikoku and Kyushu Islands are selected (Fig. 2.6), under the potential threat of tsunami disasters. I compare simulated and predicted (i.e., assimilated) waveforms at these PoIs.

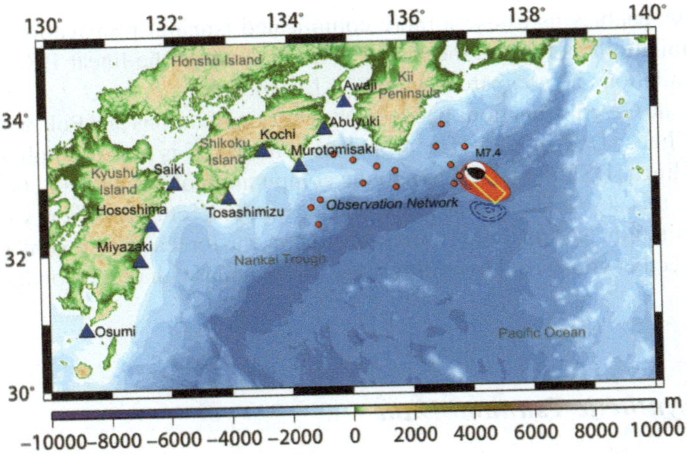

Fig. 2.6 Illustration of the observation network and near-shore Points of Interest (PoIs). The observation network in this study contains 13 DONET stations, PG1, and PG2, marked with red circles. There are nine PoIs (blue triangles) near Shikoku Island and Kyushu Island. The focal mechanism and source model are plotted according to the results of [39]. Red and blue curves are contours of uplift and subsidence, respectively. The contour interval is 0.05 m. The long axis (length) is in the NW–SE direction, whereas the short axis (width) is in the NE-SW direction. Reprinted from Wang et al. [35] Copyright 2018

2.4.3 Assimilation Setting

The bathymetry and topography dataset for computing Green's functions are derived from GEBCO_2014. I employ the finite difference method with an implicit scheme. The grid size is 30 arcsec and the time step is 1 s. The target region is 30°N–35°N, 130°E–140°E. The *JAGURS* tsunami code [3] is used to compute Green's functions for data assimilation. The same parameters of *JAGURS* used for the forward tsunami simulation are adopted. The Optimal Interpolation algorithm parameters are the same as those in the work of [18]. *JAGURS* enable the use of both the linear DSP and LLW models.

The earthquake origin time is set as $t = 0.$. The data assimilation process begins at the origin time because the tsunami reaches the network (KMC09, KMC21, and KMD13) soon after the earthquake (Fig. 2.7). Both the LLW and linear DSP models are applied. I use different assimilation time windows ranging from 2 to 24 min, at intervals of 2 min. The calculation time for the data assimilation process is less than 10 s.

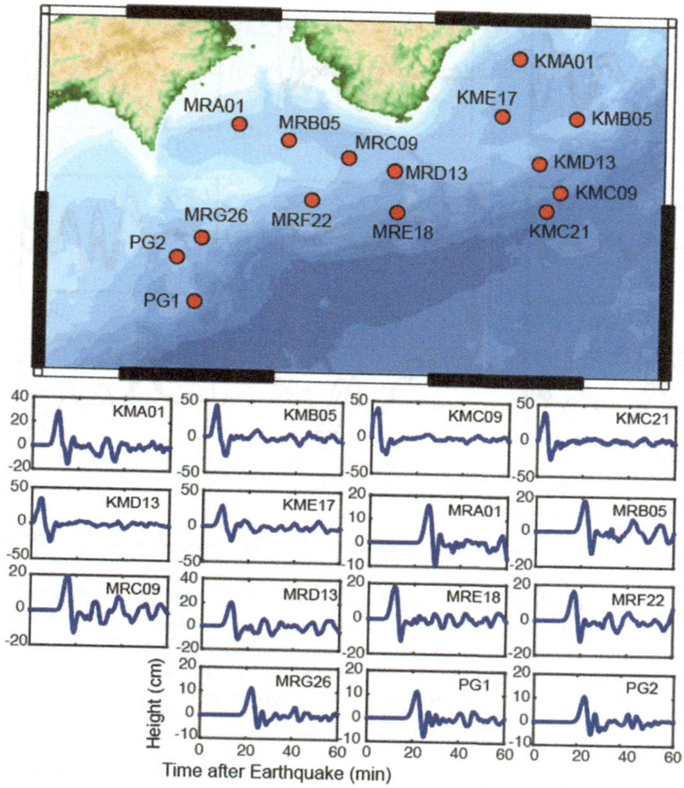

Fig. 2.7 Observational stations and waveforms of synthetic tsunamis. The tsunami reaches the KMC09, KMC21, and KMD13 stations soon after the earthquake. Therefore, the data assimilation process starts at the earthquake origin time. Reprinted from Wang et al. [35] Copyright 2018

2.4.4 *Results*

Figure 2.8 compares simulated waveforms and waveforms predicted with a 20 min assimilation time window. The waveforms forecasted using both the LLW model and the linear DSP model are consistent with with the simulated waveforms. This proves the validity of tsunami data assimilation using an observation network of DONET, PG1, and PG2. The LLW and linear DSP models have similar performances for the forecast of the first tsunami peak amplitude.

At coastal PoIs, the discrepancies of amplitude forecasted by the LLW and the linear DSP models are quite small. The main difference between two models is the tsunami arrival time. At almost every PoI, the predicted waveform using the linear DSP model has a more accurate arrival time of the first tsunami peak than that using the LLW model. The performance of two models is also related to station locations. For example, the Osumi station, located in the direction of short axis

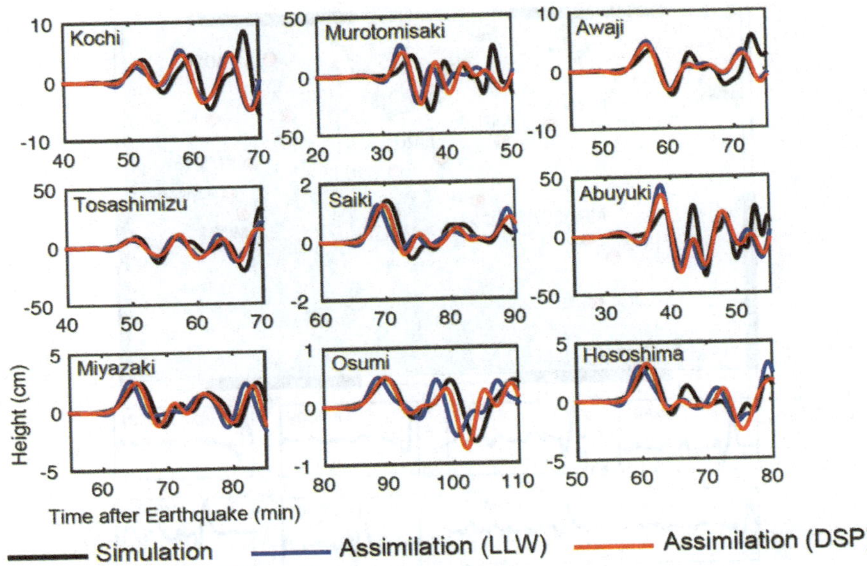

Fig. 2.8 Simulated waveforms and waveforms forecasted by data assimilation at nine near-shore PoIs. The simulated waveforms (black curves) are used as "observations" for comparison. The forecasted waveforms are calculated by Green's Function-based Tsunami Data Assimilation with the linear long-wave model (blue curves) and the linear dispersive model (red curves). The time window is 20 min. Reprinted from Wang et al. [35] Copyright 2018

(NE-SW direction), shows highly dispersive tsunami, and the linear DSP model has an evidently better performance than the LLW model. To the contrary, at Awaji or Abuyuki, which are in the long-axis direction (NW–SE direction), the predicted waveforms of two models are almost same.

The tsunami forecast accuracy [1, 9] is calculated using Eqs. (2.5) and (2.6) to assess the performance of two propagation models in data assimilation (Fig. 2.9a). The shapes of the forecast accuracy curves for both the LLW and linear DSP models are similar. Initially, the accuracies of both models are very low for a 2 min time window. At this time, the input data for assimilation is too short to provide accurate forecasting. Then, at 4 min, a sharp increase appears in the accuracy curves for both models, and the forecast accuracies exceed 85%. After that, the forecast accuracies vary slightly but generally exhibit a rising trend. The difference in forecast accuracy between the LLW model and the linear DSP model is small. Finally, after a time window of 20 min, the first tsunami peak has passed all observational stations. The forecast accuracies become saturated and stop increasing. The forecast accuracy of both models is similar, with of the linear DSP model being slightly higher.

On the other hand, the difference in arrival time of the first tsunami peak between the two models is more evident than the difference in tsunami height. I calculate the time lag as proposed by Tsushima et al. [31] to assess the forecasted arrival time quantitatively. The time lag of the ith coastal station (PoI) is defined as:

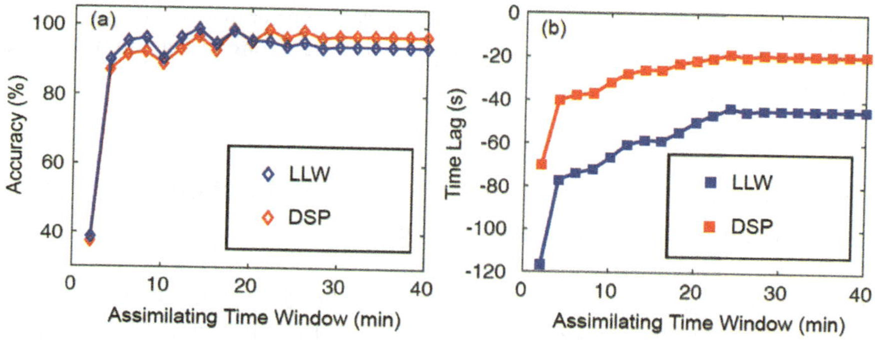

Fig. 2.9 Forecast accuracy (**a**) and time lag (**b**) of the LLW and DSP models. Reprinted from Wang et al. [35] Copyright 2018

$$\Delta T_i = t_i^S - t_i^O \tag{2.8}$$

where t_i^O is the arrival time of the maximum amplitude of the first tsunami peak at the ith station, and t_i^S is the arrival time predicted by tsunami data assimilation. A negative lag time means that the predicted arrival time is earlier than the observation. A small absolute value of time lag indicates accurate forecasting of the arrival time. The average time lag of all PoIs is calculated.

The time lags calculated by the two models are negative (Fig. 2.9b), indicating that both the LLW and linear DSP models forecast the tsunami arrival time earlier. Additionally, as the assimilation time window gradually increases, the time lag becomes closer to zero. The shape of the time lag curve is similar to that of the accuracy curve (Fig. 2.9a). From the 2 min time window to the 4 min time window, the time lag (absolute value) rapidly decreases as more input data are provided. Then, it decreases slowly. After the time window of 20 min, the discrepancy becomes very small. It is important to note that the difference between the LLW model and the linear DSP model is noticeable in the figure. The linear DSP model leads to a much smaller time lag than the LLW model, indicating that the linear DSP model performs better in forecasting the tsunami arrival time.

The results indicate that the tsunami propagation model could affect tsunami forecast accuracy of data assimilation. Two models have a similar performance for the maximum amplitude of the first tsunami peak. However, for tsunami arrival time, the linear DSP model has a much better forecast accuracy than the LLW model. The average time lag (absolute value) using the linear DSP model is relatively smaller. For individual stations, if the station is near Shikoku Island, close to the observation network, the lag time difference is not large. However, if the station is near Kyushu Island, approximately 200 km away from the observation network, the difference in lag time becomes noticeable. The limitation of long-wave approximation causes this. For the 2004 off the Kii Peninsula earthquake considered in my research, the large dip causes short-wavelength components of a tsunami, and the long-wave approximation overestimates the velocity of tsunami propagation [24]. Consequently, the tsunami arrival time predicted using the LLW model is earlier than that using the DSP model, especially for PoIs near Kyushu Island, because a long propagation distance exaggerates such errors.

2.5 Application to Real-Time Data—2015 Torishima Volcanic Tsunami Earthquake

Unusual earthquakes repeated happened 1984, 1996, 2006, and 2015 [5, 27, 30] near Torishima (Japan), which is a volcanic island along the Izu–Bonin arc. These earthquakes had a seismic magnitude of M ~ 6 but they generated anomalously large tsunamis. They were characterized as "tsunami earthquakes" [5, 25]. Different from ordinary earthquakes, the focal mechanism of these events showed a CLVD-type rather than a double-couple source. Previous studies proposed abnormal source mechanism with volcanic origins, including hydrofracturing driven by supercritical water heated by injected magma [14], ring-fault rupture [4, 29], or horizontal tensile crack opening [5]. Shuler et al. [29] found that they had longer source durations than tectonic earthquakes of similar magnitude by teleseismic body-wave analysis and examination of the frequency content. Fukao et al. [5] estimated large seafloor uplift around a submarine volcanic caldera near Torishima. They simulated tsunami recorded at OBPGs and a tide gauge successfully. Nevertheless, the complex source mechanisms of these earthquakes are still controversial, making it difficult to quantitatively relate tsunami amplitudes to seismic parameters.

On May 2, 2015 (UTC), the 2015 Torishima earthquake occurred and triggered a tsunami which was recorded on the southern coasts of Japan (Fig. 2.10). According to a JMA report, the tsunami arrived at Hachijo Island with an amplitude of 0.5 m. The amplitude at Kozu Island was 0.3 m. In addition, the tsunami was observed by OBPGs offshore Japan, including the DONET and the DSFO off Muroto Cape (Fig. 2.10; [16]). In this section, I assimilate the actual tsunami data recorded at OBPGs of the 2015 Torishima earthquake and forecast the coastal tsunami retroactively.

2.5.1 *Observed Tsunami Data*

Tsunami records observed by DONET and the DSFO off Muroto Cape are assimilated in this study, including 23 DONET stations and two DSFO stations: PG1 and PG2. The data are obtained from NIED and JAMSTEC. The sampling rate of DONET records is 0.1 s. The sampling rate of DSFO records is 1 s. The raw records are plotted in Fig. 2.11.

The raw records are processed as follows. First, I examine the records to ensure no data gaps, spikes, or repeating values in the time series. Next, I use hydrostatic relation to convert water pressure to water height. Then, I remove the tidal components by

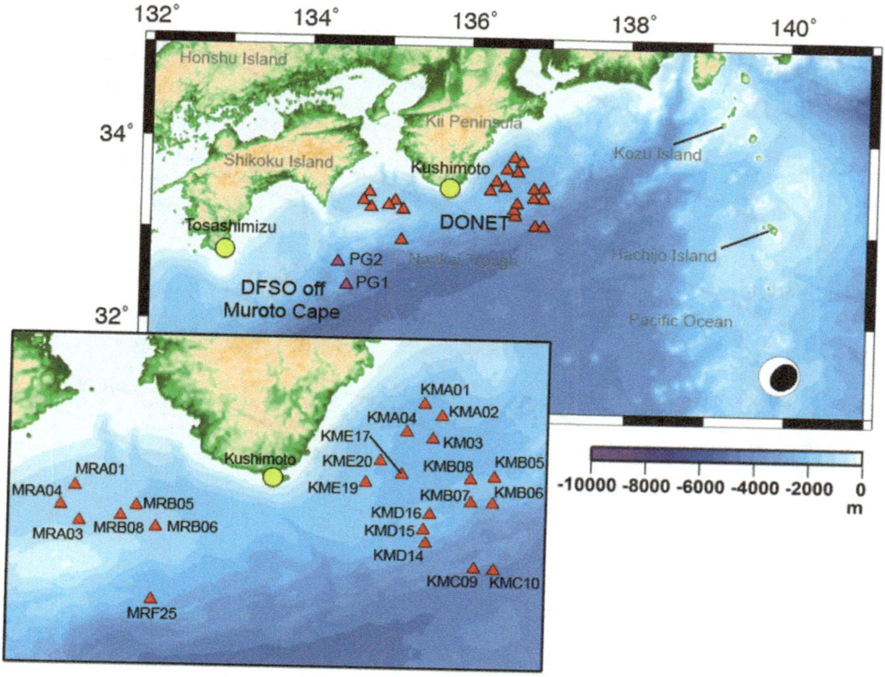

Fig. 2.10 Bathymetry map of the 2015 Torishima earthquake. Offshore observations include 23 DONET stations (red triangles) and two DSFO stations off Muroto Cape (pink triangles). The focal mechanism is based on the solution of the United States Geological Survey (USGS). Two tide gauges (i.e., Points of Interest), Kushimoto and Tosashimizu, are marked by yellow circles. Reprinted from Wang et al. [36] Copyright 2019

fitting them with trigonometric functions. Finally, I extract tsunami signals using a band-pass filter with cut-off frequencies of 0.002 and 0.01 Hz (500 and 100 s). The band-pass filter is designed same as that used by Kubota [16] for extracting tsunami signals of the 2015 and 2018 Torishima earthquakes. Because the tsunami source is not close to offshore observations, the recorded tsunami signals are not affected by the seismic waves. The processed data are plotted in Fig. 2.12.

Meanwhile, I use the data of coastal tide gauge data for comparison. Since this tsunami event is small, tsunami records were only available at two tide gauges: Kushimoto and Tosashimizu (Fig. 2.10). Kushimoto is in the Kii Peninsula, Honshu Island, whereas Tosashimizu is in Shikoku Island. Tide gauge data are obtained from the Sea Level Station Monitoring Facility, Intergovernmental Oceanographic Commission (IOC): http://www.ioc-sealevelmonitoring.org/. The sampling rate is 60 s. The data are processed in a manner similar to those of OBPGs, but the band-pass filter (0.002–0.00833 Hz) is slightly different as the Nyquist frequency of tide gauges was 0.00833 Hz.

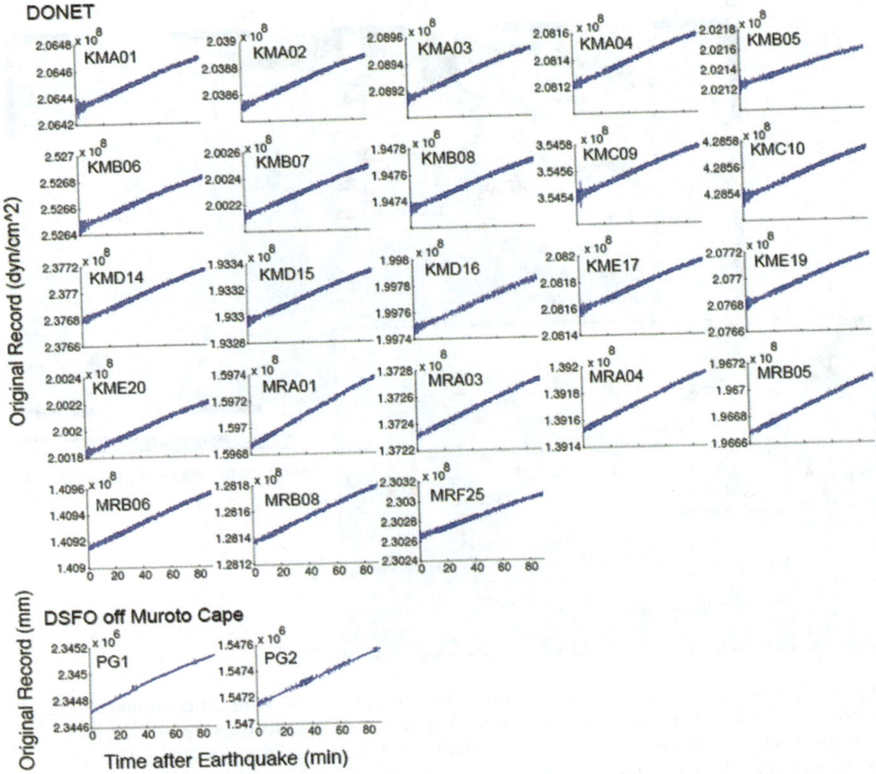

Fig. 2.11 Raw records from 25 offshore observational stations of DONET and DSFO. Reprinted from Wang et al. [36] Copyright 2019

2.5.2 *Assimilation Setting*

The tsunami arrives at the observation network (KMC10) 38 min after the earthquake. The arrival time is defined by a threshold of 0.15 cm. After that, the data assimilation process begins. The first tsunami forecast is made 40 min after the earthquake. As time passes by, the records of more stations are assimilated, and the forecasting results are continuously updated at two PoIs, Kushimoto and Tosashimizu.

The bathymetry and topography data are derived from GEBCO_2014. However, the numerical grids of two tide gauges are located on land (dry grid) when using current resolution. Therefore, I shift the station locations to the nearest wet grid. As the depth of the shifted grid of Tosashimizu exceeds 10 m, the forecasted tsunami height is corrected by the Green's Law (Eq. 1.12), assuming that the coastal depth is 1 m [2, 33]. The Green's functions are calculated by *JAGURS* model using the linear DSP model [3]. A time step of 1 s and a grid spacing of 0.5 arcmin are adopted.

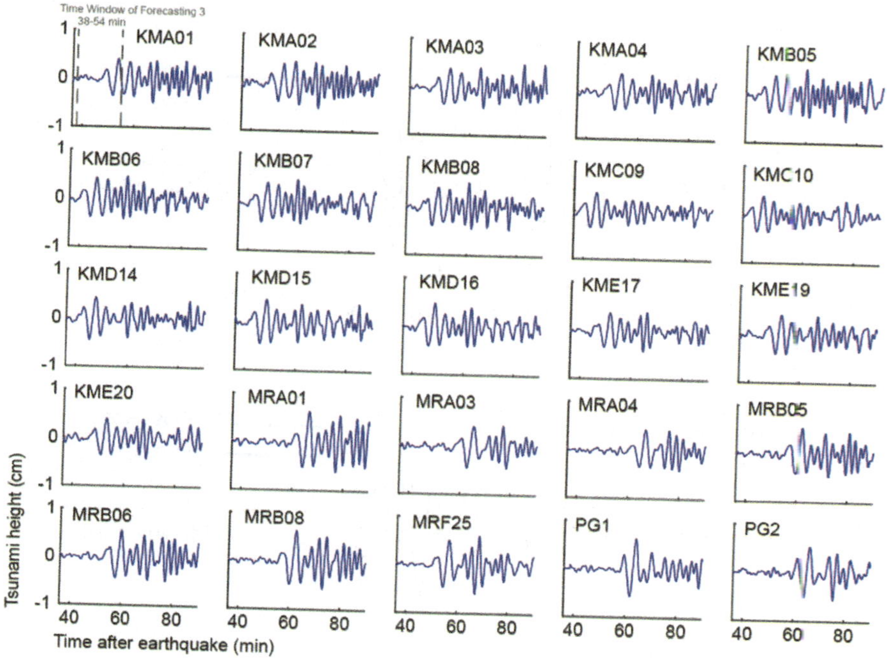

Fig. 2.12 Processed data from 25 offshore observational stations of DONET and DSFO. The time window of Forecasting 3 (38–54 min; see Sect. 2.5.2) is marked by broken lines as an example. Reprinted from Wang et al. [36] Copyright 2019

2.5.3 *Results*

The tsunami of the 2015 Torishima earthquake arrives at Kushimoto 60 min after the earthquake (Fig. 2.13a). The first-peak and maximum amplitudes are 1.3 and 2.7 cm, respectively. The forecasted waveforms with time windows of 38–40 and 38–45 min (i.e., 20 and 15 min before the arrival, respectively) are compared with observations. The former time window (38–40 min) has an acceptable forecasting performance. The forecasted arrival time is 61 min, which is slightly later than the observation. Meanwhile, the forecasted first-peak amplitude is 0.9 cm, and the forecasted maximum amplitudes is 1.4 cm. For a 38–45 min time window, the forecasted waveform is also consistent with observation. The forecasted first-peak amplitude is 1.3 cm, but the forecasted maximum amplitude is 2.2 cm.

At Tosashimizu, the first-peak amplitude is 1.2 cm, and the maximum amplitude is 6.6 cm (Fig. 2.13b). The forecasted waveforms with time windows of 38–54 min and 38–74 min (i.e., 35 and 15 min before the arrival, respectively) are compared with observations. At 54 min, data assimilation underestimates the tsunami height. Hence, it is difficult to distinguish the tsunami arrival time. For a 38–74 min time window (i.e., 15 min before the tsunami arrival) the forecasted arrival time is 89 min,

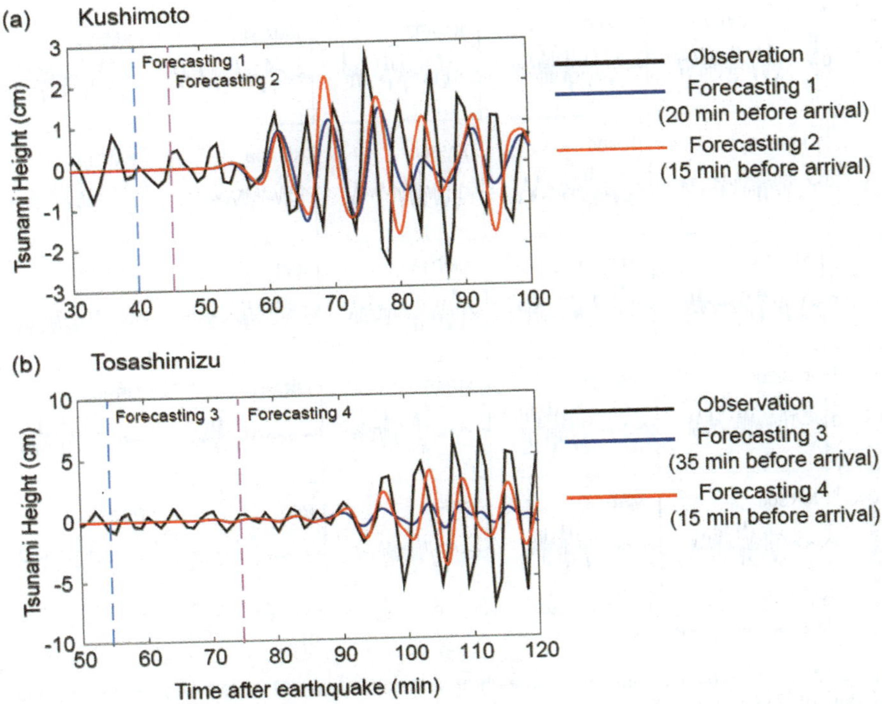

Fig. 2.13 Waveform comparison of two Points of Interest. The observed and forecasted waveforms at Kushimoto (**a**) and Tosashimizu (**b**) are compared. Black curves indicate the observation. Blue curves indicate waveforms predicted 20 min (Kushimoto) or 35 min (Tosashimizu) before tsunami arrival by data assimilation. Red curves represent waveforms predicted 15 min before tsunami arrival. Dashed blue and red lines show the different timings for making the forecast. Reprinted from Wang et al. [36] Copyright 2019

which matches quite well with observation. The forecasted first-peak and maximum amplitudes are 0.8 and 3.7 cm, respectively.

The observation and forecasted waveforms are also compared in the frequency domain (Fig. 2.14). I use the forecasted waveforms that are computed 15 min before the tsunami arrival at two stations. For Kushimoto, the spectrum of observations has an evident peak at approximately 0.0033 Hz (5.00 min) and small peaks at approximately 0.0021 Hz (7.93 min) and 0.0054 Hz (3.08 min). The spectrum of the forecasted waveform is smoother than that of observations. There is a large peak at approximately 0.0022 Hz (7.58 min) and a small peak near 0.0033 Hz (5.00 min). For Tosashimizu, the spectrum of observations has the greatest energy near 0.004 Hz (4.17 min) and another prominent peak at approximately 0.0025 Hz (6.67 min). The spectrum of forecasted waveforms generally has a similar shape, yet the amplitude is lower for some frequencies.

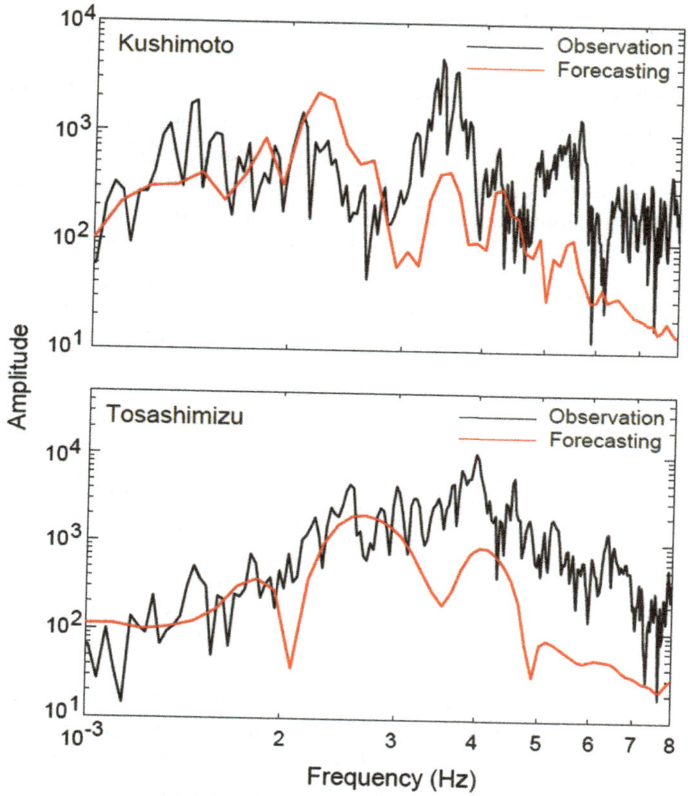

Fig. 2.14 Spectrum comparison of two Points of Interest. The spectra of observed (black curves) and forecasted (red curves) waveforms are presented. The forecasted waveforms at both two stations are computed 15 min before the tsunami arrival. Reprinted from Wang et al. [36] Copyright 2019

Although the maximum amplitude of Kushimoto is accurately predicted, the forecasting at Tosashimizu is not satisfactory. A possible reason is that Tosashimizu is inside a harbor where we lack dedicated local bathymetry. Although its amplitude has been corrected by Green's Law, the harbor effects could not be well simulated. This could also explain why the spectrum of forecasted waveforms does not match well with observations. The spectrum of tsunami waveforms largely depends on the local bathymetry [22, 23]. If the complex geometries of bays and harbors are not well simulated, discrepancies will occur. To solve this problem, bathymetric grids with higher resolution are necessary to simulate the effect of radiation damping through the mouth of the bay/harbor. Additionally, because I shift the station locations to the nearest wet grid, the arrival time predicted by data assimilation may be slightly earlier.

2.5.4 Accuracy and Number of Stations

For the 2015 Torishima volcanic tsunami earthquake, tsunami data recorded by a dense observation network are assimilated. As some stations are close to each other, the information for data assimilation may be redundant. In future operations, some stations may encounter technical problems and stop providing real-time data transmission. Hence, it is important to discuss the relationship between the forecast accuracy and the number of observational stations. Similar analyses have been proposed in previous studies (e.g., [10, 13, 19, 20]. In this section, I investigate which stations are more important for tsunami data assimilation. I use the general error function G [20] that considers the forecast ability of both tsunami amplitude and arrival time as follows.

$$G = \alpha(1 - P) + \beta Q \tag{2.9}$$

$$P = 1 - \frac{\sum (H_f^{\max} - H_o^{\max})^2 + \sum (H_f^{\text{arr}} - H_o^{\text{arr}})^2}{\sum (H_o^{\max})^2 + \sum (H_o^{\text{arr}})^2} \tag{2.10}$$

$$Q = \frac{1}{N_s} \sum \frac{\left| t_f^{\text{arr}} - t_o^{\text{arr}} \right|}{T_{\text{eva}}} \tag{2.11}$$

The subscripts f and o represent the forecasted and observed tsunamis, respectively. Equation (2.10) evaluates the forecast accuracy of tsunami amplitude, developed from the method of [31]. $H^{\max}$ and H^{arr} are the maximum and first-peak amplitude, respectively. P is a normalized index (ranging from 0 to 1). A higher (close to 1) P value means better forecasting. Using Eq. (2.11), I assess the forecast accuracy of tsunami arrival time. t^{arr} represents the arrival time of the first tsunami peak. N_s is the number of PoIs. The absolute value is calculated to avoid canceling out of the positive and negative time lags. The average time shift Q is the ratio of absolute time lag to the evacuation time, T_{eva}. A smaller Q value indicates better forecasting of arrival time. The evacuation time is defined as the interval between the forecasting time and the tsunami arrival time. The index α and β in Eq. (2.9) are two positive weight parameters which are determined as per forecasting requirements. Their relative values show the importance of the accurate forecasting of arrival time or amplitude. For simplicity, two parameters are set in such a way that $\alpha + \beta = 1.0$. As the value of G largely depends on the value of two weight parameters, I normalize the G for a fair comparison. A relatively smaller value of G suggests a higher accuracy.

The investigation starts from the entire number of observational stations. Then, the number gradually decreases. G value varies with a changing station distribution. For each station number, I search the best design of station distribution that could produce the least general error function $minG$. The gradient descent algorithm is adopted [7]. The results indicate the best station distribution of a given number of stations [20]. Hence, the importance of existing observational stations in tsunami data assimilation is calculated.

The assimilation time window of 7 min (38–45 min) is selected for Kushimoto and 36 min (38–74 min) is selected for Tosashimizu. That is to say, tsunami forecasting is made 15 min before its arrival at two PoIs. The evacuation time T_{eva} becomes 15 min.

In the first setting, two weight parameters α and β are 0.5. It indicates that I assume that the accurate forecasting of arrival time and amplitude are equally important (Fig. 2.15a). The investigation starts from the current station number 25. Initially, $minG$ increases slowly with small variations as the number of stations decreases. In this occasion, even if some "unimportant" stations are removed from the observation network, there will not be a substantial influence on the forecast accuracy as the remaining network is dense enough. This trend continues until there are ten remaining stations. After that, $minG$ increases dramatically as the number of stations decreases. In this situation, all remaining stations are significant for tsunami data assimilation. If one station encounters problems and stops working, the overall forecast accuracy will be greatly affected, making the data assimilation results no longer reliable. All in all, in the first set with $\alpha = 0.5$, $\beta = 0.5$, six is the critical number of stations. This could achieve a fair forecast accuracy ($minG = 9.37$) with the least number of stations. The observation network consists of KMC09, KMC10, MRB06, MRF25, PG1, and PG2 (Fig. 2.15b). These stations are important in tsunami data assimilation due to their critical locations. More attention should be paid to their operation and maintenance.

Furthermore, two other settings are tested: $\alpha = 0.2$, $\beta = 0.8$; $\alpha = 0.8$, $\beta = 0.2$. The former indicates a higher demand on the forecasting of arrival time as a larger β is set. In other words, it is assumed that the accurate forecasting of tsunami arrival time is more significant. To the contract, the latter indicates a higher demand on the forecast accuracy of tsunami amplitude. The value of $minG$ of these two settings are normalized for a better comparison with the first setting. These two settings have similar shapes as that of the first setting. The critical points are 6 and 16 for the second and third setting, respectively (Fig. 2.15c, d). Furthermore, the determination of "critical points" is arbitrary, which relies on the tolerance of forecasting errors in real practical. Here, the results of critical stations are calculated from a single tsunami event. To analyze the necessary design of OBPGs thoroughly, we can use Empirical Orthogonal Function analysis to find the place where the most energetic wave dynamics occur [20].

To summarize, the investigation using general error function G helps us determine the importance of offshore stations in tsunami data assimilation. More efforts could be devoted to the operation and maintenance of stations in critical locations.

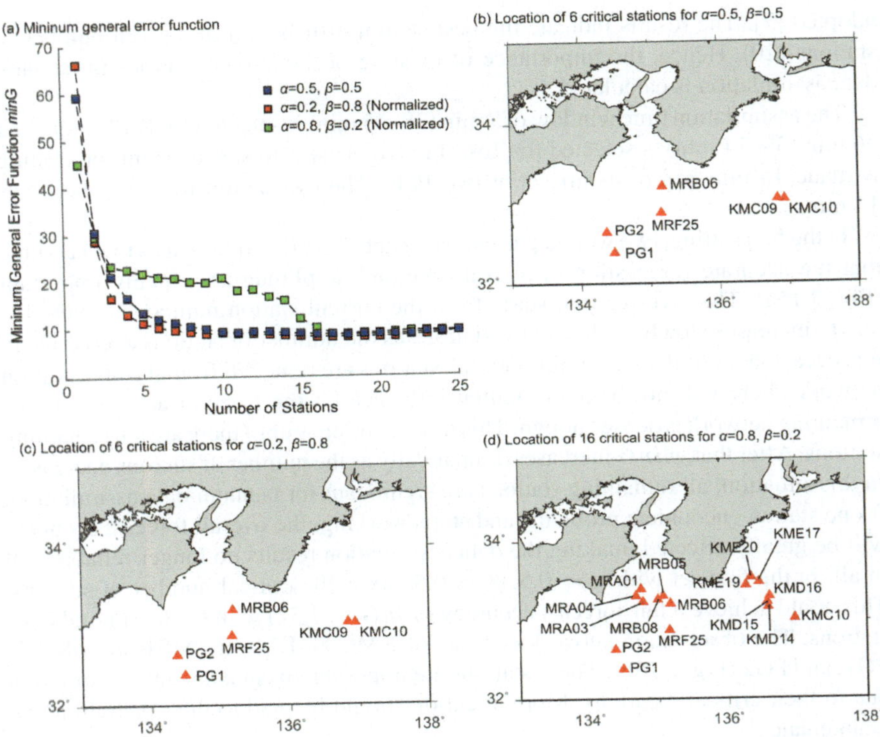

Fig. 2.15 Minimum general error function and location of critical stations. **a** Relationship between the general error function G and the number of stations. The minimum general error function $minG$ is searched by a gradient descent algorithm. The normalized results of three settings are plotted by adjusting the second and third settings to ensure that they all have same values with 25 stations. **b–d** Location of critical stations of different settings. According to the value of weight parameters α and β, the number and location of critical stations are different.

2.6 Discussion

The previous tsunami data assimilation approach [18] provides accurate tsunami forecasting for coastal regions, yet the computational cost is high. I develop a new method based on Green's functions to forecast tsunamis more rapidly. In the previous approach, the observed data are repeatedly assimilated within the time window. After that, the tsunami propagation model is run for the entire assimilated tsunami wavefield. Conversely, GFTDA does not need to calculate the whole tsunami wavefield during the data assimilation process. Instead, the waveforms at PoIs are directly synthesized using a simple matrix manipulation.

The application to the 2012 Haida Gwaii earthquake application demonstrates that GFTDA has a similar forecast accuracy as that of the previous approach, whereas the assimilation time is much shorter. Supercomputers are usually required for the assimilation process of the previous process. The high computational complexity inevitably results in time-consuming calculations. GFTDA manages to simplify the assimilation process using pre-calculated Green's functions. The assimilation process is simple and it can be achieved on personal computers. For early tsunami warnings, saving time means saving lives. The reduction of the assimilation time is critical for tsunami early warning.

The data assimilation results of the 2004 off the Kii Peninsula earthquake shows that GFTDA enables to use more complicated models in data assimilation, as long as the linearity is guaranteed. For tsunamis with strong dispersive characteristics or tsunamis that propagate over a longer distance, the linear DSP model has a better performance in forecasting the tsunami arrival time than the LLW model.

Moreover, the application to the 2015 Torishima earthquake is the first time that uses the records of a real-time observation network to forecast coastal tsunami waveforms by the data assimilation approach. The waveforms predicted 15 min before the tsunami arrival match well with real observations. Tsunamis at two coastal tide gauges, Kushimoto and Tosashimizu, are forecasted by assimilating offshore data, avoiding the complexities in the source characterization of the volcanic tsunami earthquake. Although the tsunami is small, retroactive forecasting is still meaningful regarding tsunami disaster mitigation. In case of a large tsunami event, it is appropriate to combine GFTDA with nonlinear tsunami inundation models [17]. Therefore, inundation forecasting of coastal region will be possible.

References

1. Aida I (1978) Reliability of a tsunami source model derived from fault parameter. J Phys Earth 26(1):57–73. https://doi.org/10.4294/jpe1952.26.57
2. Baba T, Hirata K, Kaneda Y (2004) Tsunami magnitudes determined from ocean-bottom pressure gauge data around Japan. Geophys Res Lett 31:L08303. https://doi.org/10.1029/2003GL019397
3. Baba T, Takahashi N, Kaneda Y, Ando K, Matsuoka D, Kato T (2015) Parallel implementation of dispersive tsunami wave modeling with a nesting algorithm for the 2011 Tohoku tsunami. Pure Appl Geophys 172:3433–3472. https://doi.org/10.1007/s00024-015-1049-2
4. Ekström G (1994) Anomalous earthquakes on volcano ring-fault structures. Earth Planet Sci Lett 128:707–712. https://doi.org/10.1016/0012-821X(94)90184-8
5. Fukao Y, Sandanbata O, Sugioka H, Ito A, Shiobara H, Watada S, Satake K (2018) Mechanism of the 2015 volcanic tsunami earthquake near Torishima, Japan. Sci Adv 4:eaao0219. https://doi.org/10.1126/sciadv.aao0219
6. Furumura T, Imai K, Maeda T (2011) A revised tsunami source model for the 1707 Hoei earthquake and simulation of tsunami inundation of Ryujin Lake, Kyushu, Japan. J Geophys Res: Solid Earth 116:B02308. https://doi.org/10.1029/2010JB007918
7. Garg R, Varna AL, Wu M (2012) An efficient gradient descent approach to secure localization in resource constrained wireless sensor networks. IEEE Trans Inf Forensics Secur 7(2):717–730. https://doi.org/10.1109/TIFS.2012.2184094
8. Gusman AR, Mulia IE, Satake K, Watada S, Heidarzadeh M, Sheehan AF (2016) Estimate of tsunami source using optimized unit sources and including dispersion effects during tsunami propagation: The 2012 Haida Gwaii earthquake. Geophys Res Lett 43:9819–9828. https://doi.org/10.1022/2016GL070140

9. Gusman AR, Sheehan AF, Satake K, Heidarzadeh M, Mulia IE, Maeda T (2016) Tsunami data assimilation of Cascadia seafloor pressure gauge records from the 2012 Haida Gwaii earthquake. Geophys Res Lett 43:4189–4196. https://doi.org/10.1002/2016GL068368
10. Heidarzadeh M, Wang Y, Satake K, Mulia IE (2019) Potential deployment of offshore bottom pressure gauges and adoption of data assimilation for tsunami warning system in the western Mediterranean Sea. Geosci Lett 6:19. https://doi.org/10.1186/s40562-019-0149-8
11. Hossen MJ, Cummins PR, Dettmer J, Baba T (2015) Time reverse imaging for far-field tsunami forecasting: 2011 Tohoku earthquake case study. Geophys Res Lett 42:9906–9915. https://doi.org/10.1002/2015GL065868
12. Hossen MJ, Cummins PR, Satake K (2017) Complete implementation of the Green's function based time reverse imaging and sensitivity analysis of reversed-time tsunami source inversion. Geophys Res Lett 44:9844–9855. https://doi.org/10.1002/2017GL074528
13. Hossen MJ, Gusman A, Satake K, Cummins PR (2018) An adjoint sensitivity method applied to time reverse imaging of tsunami source for the 2009 Samoa earthquake. Geophys Res Lett 45:627–636. https://doi.org/10.1002/2017GL076031
14. Kanamori H, Ekström G, Dziewonski A, Barker JS, Sipkin SA (1993) Seismic radiation by magma injection: an anomalous seismic event near Tori Shima, Japan. J Geophys Res: Solid Earth 98(B4):6511–6522. https://doi.org/10.1029/92JB02867
15. Korolev YP (2011) An approximate method of short-term tsunami forecast and the hindcasting of some recent events. Nat Hazard 11:3081–3091. https://doi.org/10.5194/nhess-11-3081-2011
16. Kubota T (2018) The propagation characteristic of ocean waves and tsunamis according to the analysis of pressure gauge array (In Japanese). In: The 8th conference on giant tsunami disasters, Tokyo
17. Liu PL-F, Wang X, Salisbury AJ (2009) Tsunami hazard and early warning system in South China Sea. J Asian Earth Sci 36:2–12. https://doi.org/10.1016/j.jseaes.2008.12.010
18. Maeda T, Obara K, Shinohara M, Kanazawa T, Uehira K (2015) Successive estimation of a tsunami wavefield without earthquake source data: a data assimilation approach toward real-time tsunami forecasting. Geophys Res Lett 42:7923–7932. https://doi.org/10.1002/2015GL065588
19. Mulia IE, Gusman AR, Satake K (2017) Optimal design for placements of tsunami observing systems to accurately characterize the inducing earthquake. Geophys Res Lett 44:12106–12115. https://doi.org/10.1002/2017GL075791
20. Navarrete P, Cienfuegos R, Satake K, Wang Y et al (2020) Sea surface network optimization for tsunami forecasting in the near field: application to the 2015 Illapel earthquake. Geophys J Int 221:1640–1650. https://doi.org/10.1093/gji/ggaa098
21. Okada Y (1985) Surface deformation due to shear and tensile faults in a half-space. Bull Seismol Soc Am 75(4):1135–1154
22. Rabinovich AB (1997) Spectral analysis of tsunami waves: Separation of source and topography effects. J Geophys Res: Oceans 102(C6):12663–12676. https://doi.org/10.1029/97JC00479
23. Saito T (2019) Tsunami generation and propagation. Springer Geophysics, Tokyo, pp 1–16. https://doi.org/10.1007/978-4-431-56850-6
24. Saito T, Satake K, Furumura T (2010) Tsunami waveform inversion including dispersive waves: the 2004 earthquake off Kii Peninsula, Japan. J Geophys Res: Solid Earth 115:B06303. https://doi.org/10.1029/2009JB006884
25. Sandanbata O, Watada S, Satake K, Fukao Y, Sugioka H, Ito A, Shiobara H (2018) Ray tracing for dispersive tsunamis and source amplitude estimation based on Green's law: application to the 2015 volcanic tsunami earthquake near Torishima, south of Japan. Pure Appl Geophys 175(4):1371–1385. https://doi.org/10.1007/s00024-017-1746-0
26. Satake K (1989) Inversion of tsunami waveforms for the estimation of heterogeneous fault motion of large submarine earthquakes: the 1968 Tokachi-oki and 1983 Japan Sea earthquakes. J Geophys Res: Solid Earth 94(B5):5627–5636. https://doi.org/10.1029/JB094iB05p05627
27. Satake K, Kanamori H (1991) Abnormal tsunamis caused by the June 13, 1984, Torishima, Japan, earthquake. J Geophys Res: Solid Earth 96:19933–19939. https://doi.org/10.1029/91JB01903

28. Sheehan AF, Gusman AR, Heidarzadeh M, Satake K (2015) Array observations of the 2012 Haida Gwaii tsunami using Cascadia Initiative absolute and differential seafloor pressure gauges. Seismol Res Lett 86(5):1278–1286. https://doi.org/10.1785/0220150108
29. Shuler AE, Ekström G, Nettles MK (2013) Physical mechanisms for vertical-CLVD earthquakes at active volcanoes. J Geophys Res: Solid Earth 118:1569–1586. https://doi.org/10.1002/jgrb.50131
30. Sugioka H, Fukao Y, Kanazawa T, Kanjo K (2000) Volcanic events associated with an enigmatic submarine earthquake. Geophys J Int 142(2):361–370. https://doi.org/10.1046/j.1365-246x.2000.00153.x
31. Tsushima H, Hino R, Tanioka Y, Furumura F, Fujimoto H (2012) Tsunami waveform inversion incorporating permanent seafloor deformation and its application to tsunami forecasting. J Geophys Res: Solid Earth 117:B03311. https://doi.org/10.1029/2011JB008877
32. Tsushima H, Hirata K, Hayashi Y, Tanioka Y, Kimura K et al (2011) Near-field tsunami forecasting using offshore tsunami data from the 2011 off the Pacific coast of Tohoku earthquake. Earth Planets Space 63:821–826. https://doi.org/10.5047/eps.2011.06.052
33. Wang D, Becker NC, Walsh D et al (2012) Real-time forecasting of the April 11, 2012 Sumatra tsunami. Geophys Res Lett 39(19):L19601. https://doi.org/10.1029/2012GL053081
34. Wang Y, Satake K, Maeda T, Gusman AR (2017) Green's Function-based Tsunami Data Assimilation (GFTDA): a fast data assimilation approach toward tsunami early warning. Geophys Res Lett 44:10282–10289. https://doi.org/10.1002/2017GL075307
35. Wang Y, Satake K, Maeda T, Gusman AR (2018) Data assimilation with dispersive tsunami model: a test for the Nankai Trough. Earth Planets Space 70:131. https://doi.org/10.1186/s40623-018-0905-6
36. Wang Y, Satake K, Sandabata O, Maeda T, Su H (2019) Tsunami data assimilation of cabled ocean bottom pressure records for the 2015 Torishima volcanic tsunami earthquake. J Geophys Res: Solid Earth 124:10413–10422. https://doi.org/10.1029/2019JB018056
37. Wang Y, Tsushima H, Satake K, Navarrete P (2021) Review on recent progress in near-field tsunami forecasting using offshore tsunami measurements: source inversion and data assimilation. Pure Appl Geophys 178:5109–5128. https://doi.org/10.1007/s00024-021-02910-z
38. Weatherall P, Marks KM, Jakobsson M et al (2015) A new digital bathymetric model of the world's oceans. Earth Space Sci 2(8):331–345. https://doi.org/10.1002/2015EA000107
39. Yamanaka Y (2004) EIC seismology note. http://wwweic.eri.u-tokyo.ac.jp/sanchu/Seismo_Note/2004/EIC153.html

Chapter 3
Tsunami Data Assimilation with Interpolated Virtual Stations

3.1 Linear Interpolation with Huygens–Fresnel Principle

In this chapter, I adopt virtual stations to perform tsunami data assimilation for regions without a dense observation network. The artificial waveforms of virtual stations are computed by the linear interpolation of real data. The principle of this method is similar to the Huygens–Fresnel principle. Huygens–Fresnel principle is a principle in optics [13], that every point on a wavefront is a source of spherical wavelets. The resulting amplitude at any position will be the vector sum of the amplitudes of all the individual waves of these new sources. In data assimilation, observational stations are regarded as points on the wavefront (tsunami wave). If there is a wavefront consisting of dense observational stations, the wavefield could be reproduced using the data assimilation approach. Otherwise, some information can be lost during the assimilation process. Hence, to compensate for the lost information because of sparse observations, I artificially construct waveforms at virtual stations.

Virtual station is a hypothetical concept in data assimilation. They do not really exist and are only used for building the wavefield. Figure 3.1 shows the virtual stations between two real stations (Fig. 3.1). They are located at equal interval along a straight line between two real stations. I will discuss the effects of the number of virtual stations in Sect. 3.2.3. The interpolation scheme can also be extended to three or more real stations (e.g., virtual stations inside a triangle or rectangle), but in this study I only consider two-station scenario for simplicity. The wavefront is formed by both real and virtual stations.

The "waveform" of a virtual station is constructed as the following step. First, the tsunami arrival time at a virtual station is obtained by a linear interpolation of the arrival times of two real stations:

$$t_i^{arr} = w_{iA} \cdot t_A^{arr} + w_{iB} \cdot t_B^{arr} \tag{3.1}$$

Here, the subscripts A and B represent the two neighboring real stations. t_A^{arr} and t_B^{arr} represent the arrival times of real stations defined by a threshold (e.g., 1 cm), w_{iA} and

Y. Wang, *Tsunami Data Assimilation for Early Warning*, Springer Theses,
https://doi.org/10.1007/978-981-19-7339-0_3

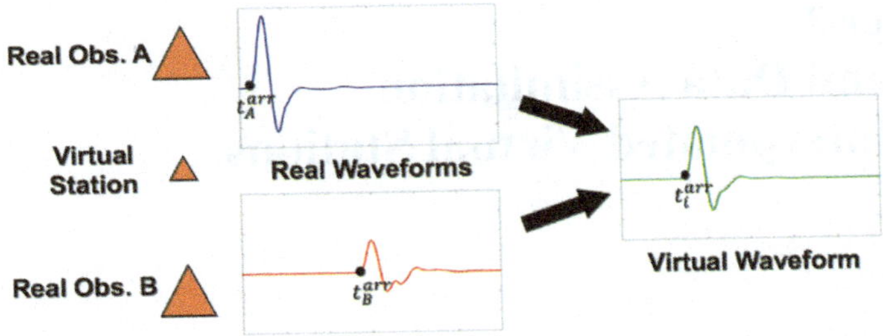

Fig. 3.1 Illustration of the linear interpolation of virtual station waveform. First, the tsunami arrival times of two real stations are determined and the arrival time of virtual station(s) is computed by weighted average. Then, the virtual waveform(s) is calculated by shifting the arrival time with a correction for water depth. Reprinted from Wang et al. [36] Copyright 2019

w_{iB} are two weight parameters, representing the relative distances between the virtual and real stations based on the assumption that the tsunami velocity around the two stations is constant. The sum of w_{iA} and w_{iB} equals one. For instance, if a virtual station is located at the middle point of two real stations, two weight parameters become 0.5.

Next, the amplitudes of two real stations are interpolated to compute the waveform of virtual stations. The water depths at different stations are considered and corrected. At real stations, the tsunami waveforms after their arrival are represented as $Y_A(t - t_A^{arr})$ and $Y_B(t - t_B^{arr})$. Then, the waveform at the i-th virtual station is computed as follows.

$$y_i(t - t_i^{arr}) = [\frac{w_{iA} \cdot Y_A(t - t_A^{arr})}{d_A{}^{-1/4}} + \frac{w_{iB} \cdot Y_B(t - t_B^{arr})}{d_B{}^{-1/4}}] \cdot d_i{}^{-1/4} \tag{3.2}$$

where d_A, d_B and d_i are water depths of two real stations and the i-th virtual station, respectively. The water depth is corrected by the Green's law that the tsunami amplitude is inversely proportional to the fourth root of water depth [29].

An evident feature of this linear interpolation method is that the virtual waveforms are calculated by shifting the arrival times considering relative distances between real and virtual stations. The arrival times are linearly interpolated by assuming a constant velocity because we lack information on the direction of wave arrival. Another feature of the interpolation method is that the correction of amplitudes relies only on the water depths at the known stations, based on the assumption that the tsunami propagates as a plane wave. I will validate my method with real data in Sect. 3.3.2. Furthermore, the virtual stations help construct a complete tsunami wavefront together with real stations. As the interpolation depends on the waveform information of two neighboring real stations, the virtual waveforms are computed after the tsunami arrives at real stations.

3.2 Test with Synthetic Data—2004 Sumatra–Andaman Earthquake

A synthetic experiment of the 2004 Sumatra–Andaman earthquake is performed to valid the improved tsunami data assimilation approach. This event occurred at 00:58:53 UTC on December 26, 2004. It triggered a tsunami resulting in over 283,000 deaths [21]. The tsunami propagated through the Bay of Bengal (Fig. 3.2) and affected India and Sri Lanka's coasts about 2 h after the earthquake [11]. Because no offshore data were available in the Bay of Bengal at the tsunami event, I use synthetic data in the synthetic experiment.

3.2.1 Synthetic Tsunami Data

The fault model is based on the results of [11]. The slip distributions are estimated the slip distributions by inverting the tide gauge and satellite altimeter data. The rupture velocity is 1.5 km/s. First, I compute the seafloor displacement using Okada's model [25] as the initial condition for tsunami propagation (Fig. 3.2). The LLW model is adopted in tsunami modeling. The bathymetry data is derived from GEBCO_2014. The computation domain is 70°E–100°E, 0°–25°N. The grid size is 2 arcmin and the time step is 1 s, which satisfies the Courant–Friedrichs–Lewy condition, necessary for a stable simulation. The simulated waveforms at six OBPGs are stored as synthetic observations: Stations 23,217, 23,218, 23,219, 23,227, 23,223, and 23,401. These stations were installed in the Bay of Bengal (north Indian Ocean) after the 2004 Sumatra–Andaman earthquake. At present, they are the only available stations for tsunami detection in this region. Then, linear interpolation (Eqs. 3.1 and 3.2) is employed to compute the waveforms at 25 virtual stations with an average interval of approximately 50 km (Fig. 3.2).

The tsunami waveforms at seven near-shore PoIs along the coasts of India, Sri Lanka, and Maldives are compared with predicted waveforms (Fig. 3.2). However, as the waveforms simulated from the source model of [11] do not match well with the real observations, probably due to inaccurate coastal bathymetry, synthetic waveforms are used for comparison. Green's functions between offshore stations (real and virtual stations) and near-shore PoIs are pre-calculated. The characteristic distance of the Optimal Interpolation algorithm is set to 20 km.

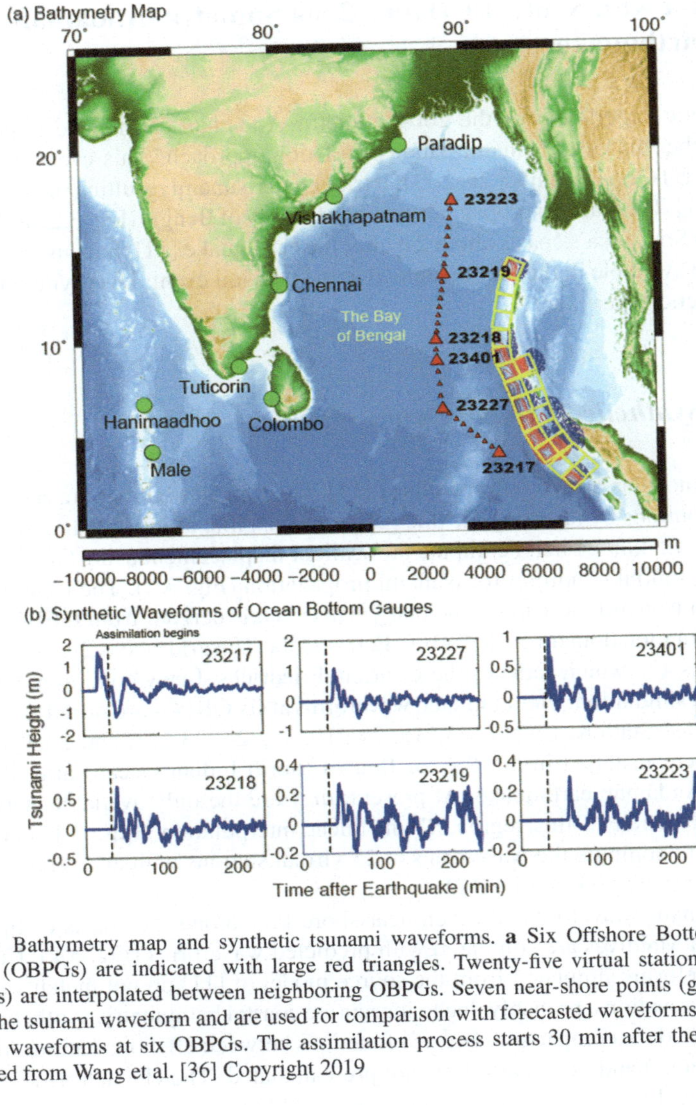

Fig. 3.2 Bathymetry map and synthetic tsunami waveforms. **a** Six Offshore Bottom Pressure Gauges (OBPGs) are indicated with large red triangles. Twenty-five virtual stations (small red triangles) are interpolated between neighboring OBPGs. Seven near-shore points (green circles) record the tsunami waveform and are used for comparison with forecasted waveforms. **b** Synthetic tsunami waveforms at six OBPGs. The assimilation process starts 30 min after the earthquake. Reprinted from Wang et al. [36] Copyright 2019

3.2.2 *Assimilation Setting and Results*

The earthquake origin time is set as $t = 0$. The synthetic tsunami arrives at OBPG Station 23,217 at approximately 30 min after the earthquake. The data assimilation process starts. The tsunami waveforms at PoIs are then forecasted by the superposition of Green's functions (Fig. 3.3).

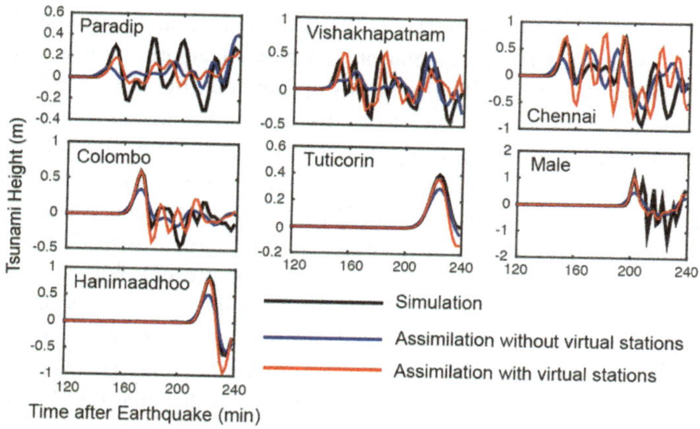

Fig. 3.3 Simulated (synthetic observation) and forecasted waveforms at seven near-shore points (i.e., Points of Interest). Black curves represent simulated waveforms used as synthetic observations. Blue curves represent waveforms forecasted without virtual stations, whereas red curves represent waveforms forecasted with virtual stations. The assimilation time window is 60 min for both cases. Reprinted from Wang et al. [36] Copyright 2019

Simulated (synthetic observation) and forecasted tsunami waveforms are compared at seven near-shore stations. The assimilation time window is 60 min. In general, the forecasted waveforms are consistent with simulated waveforms. In case of assimilation without virtual stations, the maximum amplitude of the first tsunami peak is underestimated. Conversely, the assimilation results with virtual stations have better performance, improving the forecasting of tsunami amplitude and period. For example, the simulated first-peak amplitude is 1.16 m at Male. The forecasted first-peak amplitudes without and with virtual stations are 1.06 and 0.51 m, respectively, indicating a significant improvement of prediction by including virtual stations. Overall, the forecast accuracy calculated by Eqs. (2.5) and (2.6) increases from 51.4% to 73.1% with the help of virtual stations.

In this case study, the forecast is made 90 min after the earthquake. At this time, the first tsunami peak has passed all OBPGs, resulting in an accurate forecast. Because the tsunami reaches Indian coast approximately 150 min after the earthquake and arrives at Sri Lanka and the Maldives even later, there is sufficient time for data assimilation process and issuing early warning.

3.2.3 *Effects of Interpolation Interval*

In this application, I use synthetic waveforms at six OBPGs to artificially build a dense network. Different interpolation intervals may affect forecast accuracy. Meanwhile, in the Optimal Interpolation algorithm, the covariance matrices of the forward numerical simulation and the observation are assumed to have a Gaussian correlation with a specific characteristic distance [19, 23]. In the last section, a Gaussian-distributed covariance is assumed for the correlation function of errors, with a characteristic distance of 20 km (Eq. 1.28). The relationship between the characteristic distance and the interpolation interval remains to be discussed. According to the recent study of [41], the Optimal Interpolation algorithm may have a poor performance if the interval between the neighboring stations is greater than ~50 km.

To study these effects, I use synthetic 2004 Sumatra–Andaman earthquake data and test four interpolation scenarios of virtual stations at the Bay of Bengal (Fig. 3.4). In Case 1, only four virtual stations are assumed to be present, with station intervals of ~150 km. In Case 2, 10 stations are assumed at intervals of ~100 km. Case 3 is adopted in Sect. 3.2.2, with 25 virtual stations at ~50 km intervals. In Case 4, 40 virtual stations are assumed with the smallest interval of ~30 km. For each case, I test different characteristic distances for the Optimal Interpolation algorithm, from 20 to 80 km, and calculate their forecast accuracies.

After computing the synthetic tsunamis, I calculate the forecast accuracy by Eqs. (2.5) and (2.6) for each case with different characteristic distances (Fig. 3.5). Case 1 has the lowest accuracy for all characteristic distances, and Case 2 has a better accuracy. Cases 3 and 4 have the best performance, and their results are close to each other. It can be concluded that a dense interpolation of virtual stations may help improve the forecasting up to a certain level, but beyond that, the accuracy tends to saturate. On the other hand, the number of Green's functions will also increase by adding virtual stations, which may affect the forecasting efficiency. Therefore, it is important to balance the number of virtual stations and the desired forecast accuracy.

It is also observed that the characteristics distance affects the optimum point of forecast accuracy. The smallest characteristic distance of 20 km results in the highest accuracy for Cases 3 and 4, which have interpolation intervals of ~50 and

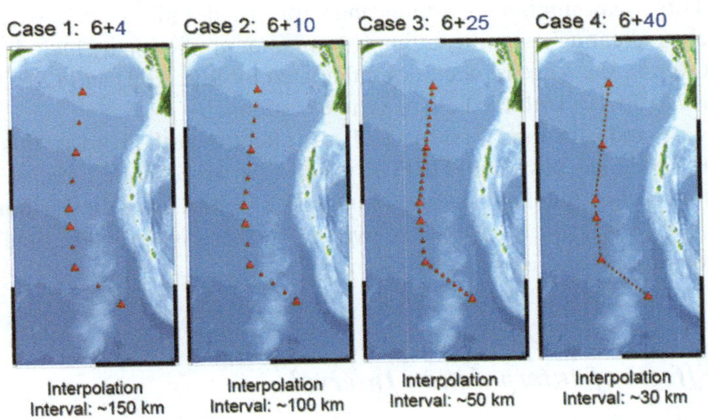

Fig. 3.4 Four interpolation scenarios of virtual stations at the Bay of Bengal. The number and location of real Offshore Bottom Pressure Gauges are fixed, but the number and interpolation interval of virtual stations are different. From Case 1 to Case 4, the interpolation becomes denser. Reprinted from Wang et al. [36] Copyright 2019

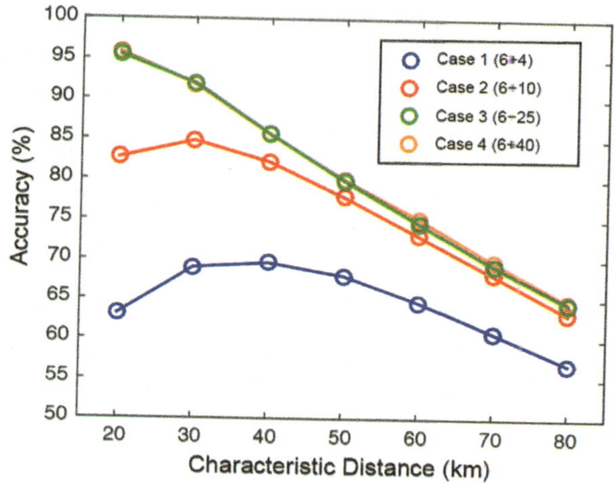

Fig. 3.5 Forecast accuracies of four scenarios versus different characteristic distances. The characteristics distances range from 20 to 80 km. The accuracy is calculated by Eqs. (2.5) and (2.6). Reprinted from Wang et al. [36] Copyright 2019

~30 km, respectively. Case 2 with an interpolation interval of ~100 km requires a 30 km characteristic distance, and Case 1 with an interpolation interval of ~150 km requires a 40 km characteristic distance to produce a high forecast accuracy.

As the interpolation becomes denser, the interval between the two stations becomes smaller. A dense interpolation is helpful to improve the forecast accuracy. In this study, the station interval should be less than ~50 km to guarantee a high accuracy over 75%. This empirical value is consistent with the distance criterion for Optimal Interpolation [41]. A characteristic distance of 20 km is appropriate in such interpolation scenario.

3.3 Test with Real Data—2009 Dusky Sound Earthquake

Another test is the 2009 Dusky Sound earthquake (M 7.8). The earthquake happened at 09:22:29 UTC on July 15, 2009, near the southwestern coast of New Zealand [6, 10, 15]. This was the largest event recorded at the Puysegur subduction zone. A tsunami was generated and recorded by coastal tide gauges around the southwestern South Island and DART stations in the south Pacific [8]. The OBS network of the Marine Observations of Anisotropy near Aotearoa (MOANA) project also recorded the tsunami [32].

3.3.1 Observed Tsunami Data

There were 30 OBSs equipped with OBPG in the west and east of South Island at the time of earthquake [32]. Raw data are available at the Ocean Bottom Seismograph

Instrument Pool (website: http://www.obsip.org). The tsunami signals are extracted by removing the tidal components and bandpass filtering at 0.0002–0.005 Hz with a fourth-order Butterworth filter. Then, the instrument responses are de-convolved, following the method of [32]. Because the tsunami was recorded by differential pressure gauges, the tsunami amplitude is further corrected by the ratio between the observed and simulated peak amplitude at each station [6]. I use these data for retroactive analysis.

I use tide gauge data for waveform comparison in order to validate the data assimilation approach. The region of interest is the western coast of South Island. There is only one tide gauge (Charleston) with available observation (Fig. 3.6a). Tsunami waveforms of tide gauges are often influenced by local bathymetry, and therefore, it is necessary to correct their amplitude based on Green's Law [4, 16, 20, 22].

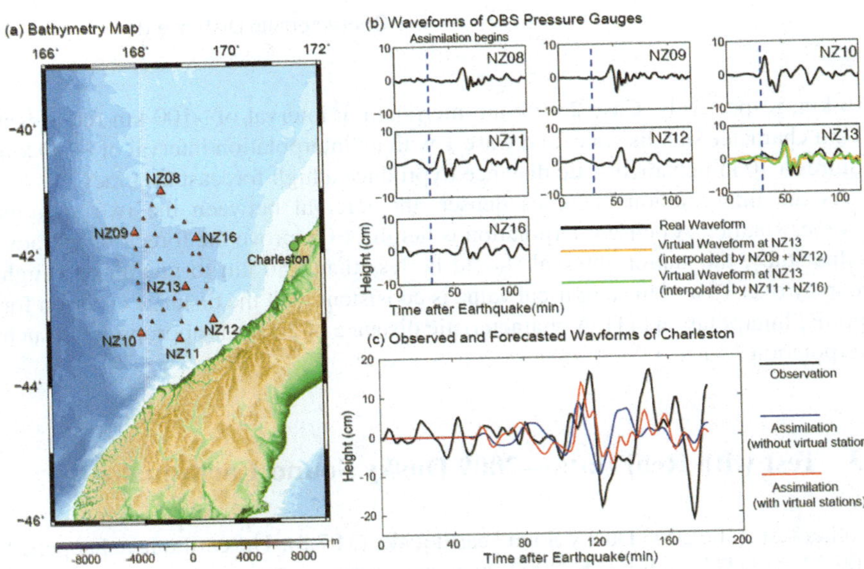

Fig. 3.6 Data assimilation of the tsunami triggered by the 2009 Dusky Sound earthquake. **a** Offshore Bottom Pressure Gauges (OBPGs) and Charleston tide gauge. There are seven real OBPGs (large red triangles) and 28 virtual stations (small red triangles) used for data assimilation. **b** Observed tsunami waveforms at seven OBPGs. Real (black curve) and virtual (yellow and green curves) waveforms at NZ13 are compared. **c** Observed and forecasted waveforms at Charleston. The forecast is made with a time window of 40 min. Modified from Wang et al. [36] Copyright 2019

3.3.2 Assimilation Setting

Although 30 OBSs were installed during the MOANA project, data from only a few stations are available for data assimilation. Among the stations in the west, NZ15 was trawled up by a fishing vessel, and NZ17 has not been recovered [42]. NZ01 and NZ02 waveforms were clipped [32]. Additionally, some stations are too far from Charleston. The tsunami of the 2009 Dusky Sound earthquake reached these stations even later than tide gauges. Hence, there are only seven stations for data assimilation: NZ08, NZ09, NZ10, NZ11, NZ12, NZ13, and NZ16.

Twenty-eight virtual stations are interpolated between real stations using Eqs. 3.1 and 3.2. The average interpolation interval is approximately 50 km. I only choose adjacent station pairs (e.g., NZ10 and NZ11) so that virtual stations do not overlap with each other. To validate the linear interpolation method, I use NZ13 as an example, which is nearly located in the line between NZ09 and NZ12 and the line between NZ11 and NZ16. I interpolate a virtual waveform at NZ13 using data from NZ09 and NZ12 (yellow curve in Fig. 3.6b) and another virtual waveform at NZ13 using the data from NZ11 and NZ16 (green curve in Fig. 3.6b). Then I compare virtual waveforms with real waveform (black curve).

As for the setting of data assimilation, I use a characteristic distance of 20 km for the Optimal Interpolation. The bathymetry grid data is derived from GEBCO_2014. The computation domain is 163°E–175°E, 49°S–37°S. The grid size is 0.5 arcmin, and the time step is 1 s. The waveforms of both real and virtual stations are used as the input data to overcome the problem of sparse observations.

3.3.3 Results

In the test of NZ13, the arrival time and amplitude of the first tsunami peak are quite consistent between real observation and two virtual waveforms (Fig. 3.6b). Hence, it validates the linear interpolation method of virtual stations. Although there are slight discrepancies, interpolated waveforms at virtual stations could supplement insufficient real observations in tsunami data assimilation and compensate for the lost information due to sparse stations.

The tsunami of the 2009 Dusky Sound earthquake arrives at the first real station (NZ10) 30 min after the earthquake. Then, the assimilation process begins. I compare observed and forecasted waveforms at Charleston coastal tide gauge (Fig. 3.6c). The time window is 40 min, which means that the prediction is made 40 min after the tsunami arrival at NZ10 or approximately 30 min before the arrival at Charleston. The forecasted waveform is generally consistent with tide gauge observation in terms of arrival time (phase). Nevertheless, the amplitude of the waveform forecasted without virtual stations (8.4 cm) is evidently smaller than that of the observation (16.0 cm). The accuracy is 52.5% [1]. Conversely, the application of virtual stations leads to a better forecast of tsunami amplitude (13.4 cm). The accuracy is 83.8%.

3.4 Application to Far-Field Event—2015 Illapel Earthquake

The 2015 Illapel earthquake (M 8.3) occurred offshore Illapel (Chile) on September 16, 2015, at 22:54:33 (UTC). The earthquake is a result of the thrust faulting on the interface between the Nazca and South America Plates. The epicenter is 31.573°S, 71.674°W, at a depth of 22.4 km. A tsunami was generated, which arrived at the Chilean coastal region within 8 min after the earthquake [2, 3] and resulted in flooding in coastal cities. The tsunami also propagated through the Pacific Ocean although its amplitude was smaller than that of the 2010 and 2014 tsunamis [30]. DART tsunameters in the Pacific Ocean recorded the tsunami signals. These devices are able to provide real-time tsunami time series at the open ocean, which can be used to reconstruct the earthquake source model [17, 39]. Ren et al. [28] studied the features of tsunami waves in the deep ocean by using the records of 16 DART tsunameters of the 2015 Illapel earthquake.

In this research, I use data assimilation approach to forecast far-field tsunami of earthquakes in Chile. Chilean earthquakes can trigger tsunamis affecting not only the South American coast but also the regions far away from the source, including Easter Island [9], Oceania [14], and even Japan [31]. DART tsunameters off the Chilean coast enables the application of the tsunami data assimilation approach. The DART stations are 32,401 and 32,402 owned and maintained by the Hydrographic and Oceanographic Service of the Chilean Navy (Servicio Hidrográfico y Oceanográfico de la Armada de Chile in Spanish), and stations 32,403, 32,404, and 34,420 owned and maintained by the Cooperative Effort DART 4G Buoy (Fig. 3.7a). These stations record water levels and transmit signals in real time [12]. Hence, the 2015 Illapel earthquake is adopted as a case study of far-field tsunami data assimilation. I use tsunami data assimilation approach to retroactively forecast the tsunami in the east Pacific.

3.4.1 Observed and Synthetic Tsunami Data

The tsunami data used in this study include the records from DART tsunameters and tide gauges. The records of four DART tsunameters are used as the input for data assimilation. The records of tide gauge and other DART tsunameters are used to compare with forecasted waveforms.

The DART records of the 2015 Illapel earthquake are available at NOAA's website (https://www.ngdc.noaa.gov/hazard/dart/2015chile.html). The raw data has already been processed by the Tsunami Detection Algorithm to extract tsunami signals [24]. The tsunami amplitude is computed by subtracting predicted water pressures from the observations to remove the effects of tides and low frequency fluctuations. Additionally, I remove high-frequency components such seismic waves by applying a low-pass filter (cut-off frequency: 0.002 Hz) and a zero mask before the tsunami

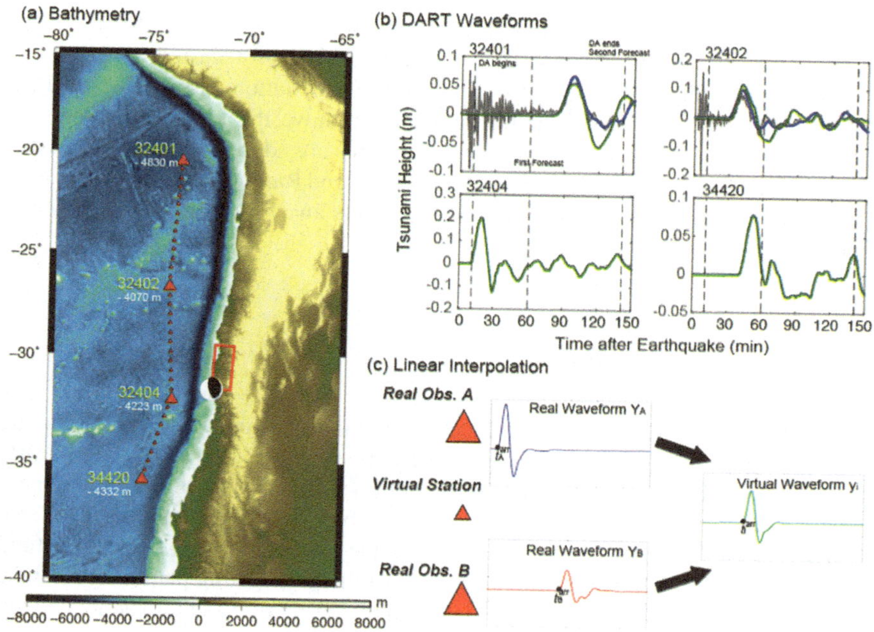

Fig. 3.7 Tsunami data assimilation of the 2015 Illapel earthquake. **a** Bathymetry and DART tsunameters. The data at four DART tsunameters and 32 virtual stations are adopted in tsunami data assimilation. The focal mechanism is plotted using the USGS solution. The source model of [28] is plotted with a red rectangle. **b** Waveforms at DART stations. Grey curves represent the unfiltered data directly obtained from NOAA's website, blue curves are observations (i.e., waveforms after filtering), and green curves stand for synthetic waveforms. Observations at 32,401 and 32,402 and synthetic waveforms at 32,404 and 34,420 are used as the input for data assimilation. The time for the beginning and end of data assimilation process are marked by dashed lines. **c** Linear interpolation for computing waveforms at virtual stations. Reprinted from Wang et al. [37] Copyright 2019

arrival. The reason is that only the waveforms after the tsunami arrival are necessary for data assimilation. For most tsunamis generated by mega-thrust earthquakes like the 2015 Illapel event, OBPGs record tsunamis whose first peak is free of harbor effects and coastal reflections [38].

However, at the time of this event, there were only two DART stations (32,401 and 32,402) in operation. Therefore, besides observed waveforms of these two stations (blue curves), I artificially compute synthetic waveforms (green curves) at stations 32,404 and 34,420 (Fig. 3.7b) by using the source model of [28] as a supplement. The station 32,403 is skipped because it is close to station 32,401. The fault model has a focal depth of 25 km. The strike, dip, and rake angles are 4°, 19°, and 90°, respectively. The fault size (length × width) is 230 km × 100 km. A uniform slip of 4.64 m is adopted. The tsunami waveforms simulated from this model have a high consistency with real observations at near-field and far-field DART tsunameters [28].

Furthermore, the records of DART station 32,412, which is located off the southwest of Peru, are also adopted in this study to compare with the forecasted waveform.

Tide gauge data are obtained from the Sea Level Station Monitoring Facility, IOC (http://www.ioc-sealevelmonitoring.org/). I remove the tidal components by polynomial fitting. The records of seven tide gauges are adopted: San Felix, Easter, Rikitea, Nuku Hiva Island, Papeete Tahiti, Huahine, and Rarotonga (Fig. 3.8a). These tide gauges are hundreds to ten thousand kilometers away from the South America continent.

3.4.2 Assimilation Setting

Four DART tsunameters are adopted for data assimilation, among which I interpolate 32 virtual stations (Fig. 3.7a, b). Green's functions are pre-calculated by the *JAGURS* tsunami code [5]. I use the linear DSP model on spherical coordinates. Nonlinear effects (e.g., bottom friction) are neglected. The bathymetry data are derived from GEBCO_2014. The grid size is 2 arcmin, and the time step is 1 s. For the far-field tsunamis of the 2015 Illapel earthquake, most of the propagating path is in the wide Pacific Ocean. Nevertheless, the exact location of coastal tide gauges cannot be well represented under current bathymetry resolution. The corresponding grid points of tide gauges sometimes become dry grids (on the land). To avoid this problem, I shift the location of the tide gauge to the nearest wet grid.

Tsunami waveforms at eight PoIs, including DART tsunameters and tide gauges, are predicted by data assimilation (Fig. 3.8a). Compared to DART tsunameters, tide gauges are more affected by local topography. They usually have larger amplitudes and more complicated spectrums than stations in the open ocean [26, 27].

3.4.3 Results

The earthquake origin time is defined as $t = 0$. The tsunami arrival time is defined by a threshold of 0.02 m. The tsunami arrives at station 32,404 at $t = 12$ min, after which the data assimilation process starts. As a limitation of data assimilation with virtual stations, the waveforms at virtual stations are not computed until the tsunami arrives at both neighboring real stations. This may cause some delay in forecasting, but concerning the propagation time of far-field tsunamis, such delay is not important. The first tsunami forecasting is made at $t = 1$ h when the first tsunami peak has passed stations 32,402 and 32,404, as well as virtual stations interpolated between them. Then, as time passes, more data are assimilated to improve the forecast accuracy. The assimilation process ends at $t = 2$ h 20 min. At this time, the first tsunami peak has passed four DART tsunameters.

I compare forecasted and observed tsunami waveforms at PoIs (Fig. 3.8b). The tsunami heights at tide gauges have already been corrected by Green's Law, assuming

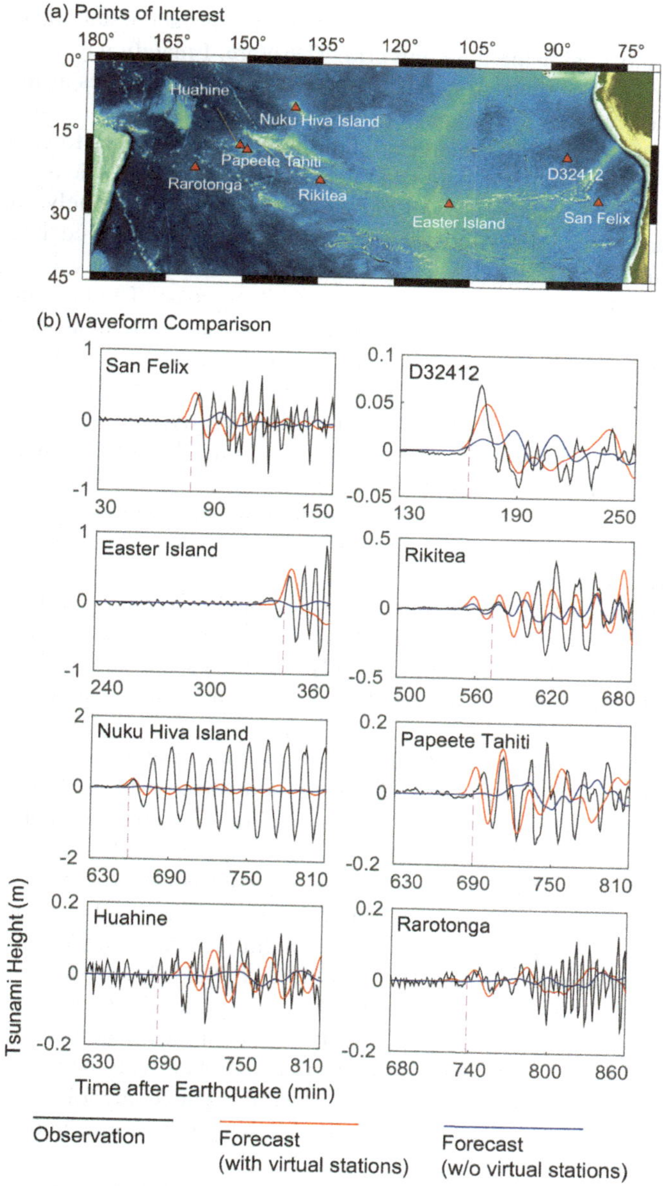

Fig. 3.8 Locations of and waveforms at Points of Interest (PoIs). **a** Locations of eight PoIs. The far-field tsunami in the east Pacific is forecasted from the ocean off the Chilean coast to regions in Oceania. **b** Observed and forecasted waveforms at eight PoIs. Black curves represent observed waveforms. Red curves stand for forecasted waveforms by data assimilation with virtual stations. Blue curves stand for forecasted waveforms by data assimilation without virtual stations. Modified from Wang et al. [37] Copyright 2019

that the coastal depth is ~1 m [4, 33, 35]. The waveforms at coastal tide gauges may not be fully modeled using 2 arcmin grid. Therefore, I mainly pay attention to the amplitude and arrival times of the first tsunami peak at tide gauges as the first peak is less affected by local topography or harbor effects [7, 40].

Using only four DART tsunameters for data assimilation, the forecasted waveforms (blue curves) do not match with the observations at PoIs. The tsunami amplitude is greatly underestimated, and the arrival time is not accurately predicted. In contrast, data assimilation with virtual stations (red curves) improved the forecasting results substantially (Fig. 3.8).

San Felix tide gauge is located near the Chilean coast. The tsunami arrives at this station 1 h 17 min after the 2015 Illapel earthquake. The first-peak amplitude is 0.39 m. The arrival time predicted by data assimilation is 1 h 13 min after the earthquake, and the forecasted amplitude is 0.41 m. The observed and forecasted waveforms have a high consistency. As the tsunami arrives at San Felix before the end of the data assimilation process (t = 2 h 20 min), I use the results forecasted at t = 1 h 6 min. DART station 32,412 is the only PoI in the open ocean. The tsunami arrives at 2 h 45 min after the earthquake with an amplitude of 0.07 m. Data assimilation with virtual stations successfully forecasts the arrival time (2 h 45 min) and first-peak amplitude (0.05 m).

The tsunami reached Easter Island tide gauge at 5 h 36 min after the earthquake, and its first-peak amplitude is 0.40 m. Data assimilation forecasts the tsunami arrival time (5 h 34 min) accurately. Additionally, the forecasted first-peak amplitude is 0.51 m, higher than the first observed peak, but at the same level as the following observed peaks.

Tide gauges in Oceania are far from the epicenter, and it takes a long time for tsunami propagation. Despite this, the forecasted waveforms are generally consistent with observations. At most PoIs, the arrival times of the first tsunami peak are fairly forecasted, including Nuku Hiva Island, Papeete Tahiti, and Rarotonga. At Rikitea, the forecasted arrival time is approximately 18 min earlier than the observed value. At Huahine, the predicted tsunami arrives approximately 14 min later than real observation. Regarding the first-peak amplitude, data assimilation with virtual stations underestimates the amplitude at Rikitea, Nuku Hiva Island, and Papeete Tahiti, while at Huahine and Rarotonga, the first-peak amplitude is slightly overestimated. In this study, although I mainly pay attention to the first tsunami peak, the following waveforms after the first peak match well with the real observations at most stations. However, at Nuku Hiva Island tide gauge, the following waveform demonstrates strong oscillation characteristics, which are not captured well by the forecasted waveform.

Overall, the forecast accuracy is 87.5% considering all PoIs as per Eqs. (2.5) and (2.6). The average value of absolute time lag is 5.75 min using Eq. (2.8). Considering the long travel time of several hours, the arrival time forecasting is judged to be accurate. The forecasted and observed arrival times are compared at each PoI (Fig. 3.9). The slope of the regression line is 1.0037, with an R-square value of 0.999. Hence, tsunami data assimilation does not have a bias in forecasting the arrival time.

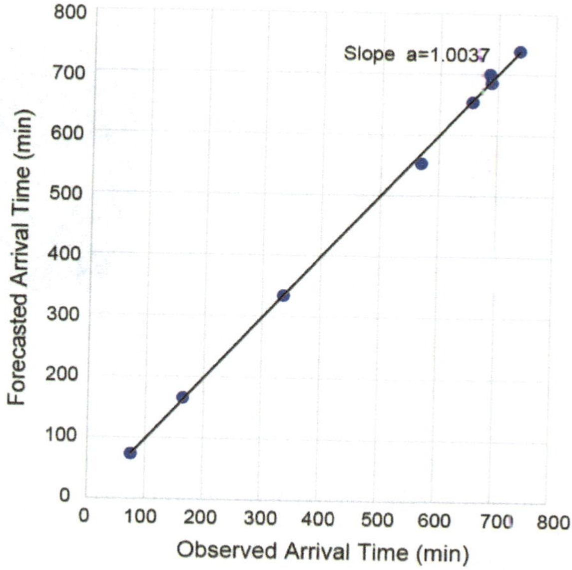

Fig. 3.9 Observed and forecasted tsunami arrival time at eight Points of Interest. Reprinted from Wang et al. [37] Copyright 2019

3.5 Discussion

The adoption of virtual stations enables us to perform tsunami data assimilation in regions without a dense observation network, such as the Bay of Bengal. The new method does not introduce any new information except the water depth. It is based on the assumption that a tsunami propagates as a plane wave, and that the wavefront is a nearly straight line. Hence, the tsunami amplitude in the open sea depends on the water depth in most cases. The data assimilation approach resembles the Huygens–Fresnel Principle. If there is no dense observation network, the previous assimilation procedure can lose some information of the wave as there are not enough "secondary sources" on the wavefront. Therefore, the interpolated virtual stations are an approximation of the lost information because of sparse stations. They help reconstruct the wavefront and prevent information loss to some extent. Of course, if there is enough budget, it will be better to build sufficient real stations to prevent such information loss instead of using interpolated virtual stations as an approximation.

As this method is based on the assumption that the tsunami travels as a plane wave, I acknowledge that it will have some limitations when this assumption cannot be satisfied. Generally, there are mainly two scenarios where the method of virtual stations that cannot be applied. In the first scenario, the source is close to the observational stations (Fig. 3.10a), and the tsunami wavefront is more like a spherical wave instead of a plane wave. Therefore, it is not applicable to create waveforms at virtual stations. However, Tanioka [34] proposed a tsunami data assimilation method based on OBPG data in the earthquake source region. Inoue et al. [18] successfully applied this method to synthetic studies of the 1952 Tokachi-oki earthquake (M 8.2)

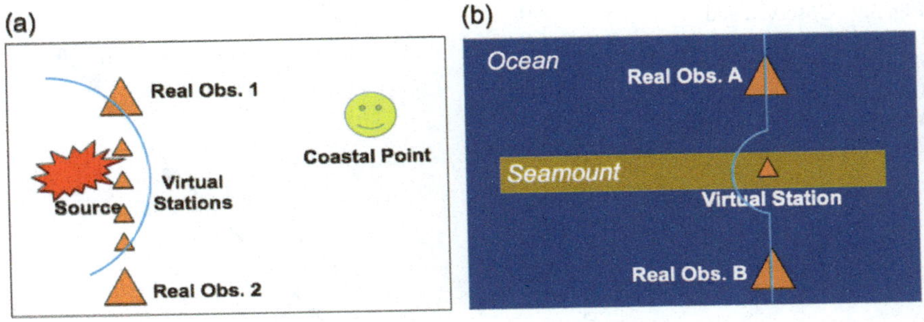

Fig. 3.10 Two scenarios where the virtual-station method cannot be adopted. **a** The scenario where the source is close to the observational stations (marked by orange triangles). The blue curve indicates the tsunami wavefront. **b** The scenario where the bathymetry varies very quickly

and the 1968 Tokachi-oki earthquake (M 8.1). Although this method is different from the Optimal Interpolation algorithm, it can help in cases where the virtual-station method cannot be applied. In the second scenario, the bathymetry varies very quickly. For example, when there is a seamount between two real observational stations (Fig. 3.10b), the tsunami wavefront will be bent and no longer be a straight light. Because the tsunami velocity depends on water depth, the bathymetry change should be smooth to satisfy the plane-wave assumption. Except for these two scenarios, the virtual waveforms calculated by linear interpolation could compensate for the lack of real observations, which has been verified by the case study on the 2009 Dusky Sound earthquake. Hence, it can improve tsunami data assimilation performance and can make it affordable for more countries and regions.

References

1. Aida I (1978) Reliability of a tsunami source model derived from fault parameter. J Phys Earth 26(1):57–73. https://doi.org/10.4294/jpe1952.26.57
2. An C, Meng L (2017) Time reversal imaging of the 2015 Illapel tsunami source. Geophys Res Lett 44:1732–1739. https://doi.org/10.1002/2016GL071304
3. Aránguiz R, González G, González J et al (2016) The 16 September 2015 Chile tsunami from the post-tsunami survey and numerical modeling perspectives. Pure Appl Geophys 173:333–348. https://doi.org/10.1007/s00024-015-1225-4
4. Baba T, Hirata K, Kaneda Y (2004) Tsunami magnitudes determined from ocean-bottom pressure gauge data around Japan. Geophys Res Lett 31:L08303. https://doi.org/10.1029/2003GL019397
5. Baba T, Takahashi N, Kaneda Y, Ando K, Matsuoka D, Kato T (2015) Parallel implementation of dispersive tsunami wave modeling with a nesting algorithm for the 2011 Tohoku tsunami. Pure Appl Geophys 172:3433–3472. https://doi.org/10.1007/s00024-015-1049-2
6. Beavan J, Samsonov S, Denys P, Sutherland R, Palmer N, Denham M (2010) Oblique slip on the Puysegur subduction interface in the 2009 July MW 7.8 dusky sound earthquake from GPS and InSAR observations: implications for the tectonics of southwestern New Zealand. Geophys J Int 183:1265–1286. https://doi.org/10.1111/j.1365-246X.2010.04798.x

7. Calisto I, Miller M, Constanzo I (2016) Comparison between tsunami signals generated by different source models and the observed data of the Illapel 2015 earthquake. Pure Appl Geophys 173:1051–1061. https://doi.org/10.1007/s00024-016-1253-8
8. Clark KJ, Johnson PN, Turnbull IM, Litchfield NJ (2011) The 2009 Mw 7.8 earthquake on the Puysegur subduction zone produced minimal geological effects around dusky sound, New Zealand. NZ J Geol Geophys 54:237–247. https://doi.org/10.1080/00288306.2010.543690
9. Fritz HM, Petroff CM, Catalán PA et al (2011) Field survey of the 27 February 2010 Chile tsunami. Pure Appl Geophys 168:1989–2010. https://doi.org/10.1007/s00024-011-0283-5
10. Fry B, Bannister S, Beavan J, Bland L et al, GeoNet team (2010) The Mw 7.6 dusky sound earthquake of 2009: Preliminary report. Bull NZ Soc Earthq Eng 43(1):24–40. https://doi.org/10.5459/bnzsee.43.1.24-40
11. Fujii Y, Satake K (2007) Tsunami source of the 2004 Sumatra-Andaman earthquake inferred from tide gauge and satellite data. Bull Seismol Soc Am 97(1A):S192–S207. https://doi.org/10.1785/0120050613
12. González FI, Bernard EN, Meinig C et al (2005) The NTHMP tsunameter network. Nat Hazards 35:25–39. https://doi.org/10.1007/s11069-004-2402-4
13. Hadamard J (1924) Le principe de Huygens (In French). Bull Soc Math Fr 52:610–640
14. Hébert H, Heinrich P, Schindelé F, Piatanesi A (2001) Far-field simulation of tsunami propagation in the Pacific Ocean: impact on the Marquesas Islands (French Polynesia). J Geophys Res: Oceans 106:9161–9177. https://doi.org/10.1029/2000JC000552
15. Heidarzadeh M, Gusman AR (2019) Application of dense offshore tsunami observations from Ocean Bottom Pressure Gauges (OBPGs) for tsunami research and early warnings. In: Geological disaster monitoring based on sensor networks. Springer, Natural Hazards. Springer, Singapore. https://doi.org/10.1007/978-981-13-0992-2_2
16. Heidarzadeh M, Harada T, Satake K, Ishibe T, Gusman AR (2016) Comparative study of two tsunamigenic earthquakes in the Solomon Islands: 2015 Mw 7.0 normal-fault and 2013 Santa Cruz Mw 8.0 megathrust earthquakes. Geophys Res Lett 43(9):4340–4349. https://doi.org/10.1002/2016GL068601
17. Heidarzadeh M, Murotani S, Satake K, Ishibe T, Gusman AR (2016) Source model of the 16 September 2015 Illapel, Chile, Mw 8.4 earthquake based on teleseismic and tsunami data. Geophys Res Lett 43(2):643–650. https://doi.org/10.1002/2015GL067297
18. Inoue M, Tanioka Y, Yamanaka Y (2019) Method for near-real time estimation of tsunami sources using ocean bottom pressure sensor network (S-Net). Geosciences 9(7):310. https://doi.org/10.3390/geosciences9070310
19. Kalnay E (2003) Atmospheric modelling, data assimilation and predictability. Cambridge University Press, Cambridge. https://doi.org/10.1017/CBO9780511802270
20. Kontar YA, Santiago-Fandiño V, Takahashi T (2013) Tsunami events and lessons learned: environmental and societal significance. Springer, Berlin. https://doi.org/10.1007/978-94-007-7269-4
21. Lay T, Kanamori H, Ammon CJ, Nettles M et al (2005) The great Sumatra-Andaman earthquake of 26 December 2004. Science 308:1127–1133. https://doi.org/10.1126/science.1112250
22. Leonard R (2006) Analysis of tide gauge records from the December 2004 Indian Ocean tsunami. Geophys Res Lett 33:L17602. https://doi.org/10.1029/2006GL026552
23. Maeda T, Obara K, Shinohara M, Kanazawa T, Uehira K (2015) Successive estimation of a tsunami wavefield without earthquake source data: a data assimilation approach toward real-time tsunami forecasting. Geophys Res Lett 42:7923–7932. https://doi.org/10.1002/2015GL065588
24. Meinig C, Stalin SE, Nakamura AI, González F, Milburn HB (2005) Technology developments in real-time tsunami measuring, monitoring and forecasting. In: Proceedings of OCEANS 2005 MTS/IEEE, vol 2, pp 1673–1679. https://doi.org/10.1109/OCEANS.2005.1639996.
25. Okada Y (1985) Surface deformation due to shear and tensile faults in a half-space. Bull Seismol Soc Am 75(4):1135–1154
26. Rabinovich AB (1997) Spectral analysis of tsunami waves: Separation of source and topography effects. J Geophys Res: Oceans 102(C6):12663–12676. https://doi.org/10.1029/97JC00479

27. Rabinovich AB, Eblé MC (2015) Deep-ocean measurements of tsunami waves. Pure Appl Geophys 172:3281–3312. https://doi.org/10.1007/s00024-015-1058-1
28. Ren Z, Yuan Y, Wang P et al (2017) The September 16, 2015 Mw 8.3 Illapel, Chile earthquake: characteristic of tsunami wave from near-field to far-field. Acta Oceanol Sin 36:73–82. https://doi.org/10.1007/s13131-017-1005-3
29. Satake K (2015) Treatise on geophysics, vol 4, 2nd edn. Elsevier, Oxford, pp 447–504
30. Satake K, Heidarzadeh M (2017) A review of source models of the 2015 Illapel, Chile earthquake and insights from tsunami data. Pure Appl Geophys 174:1–9. https://doi.org/10.1007/s00024-016-1450-5
31. Satake K, Heidarzadeh M, Qurioz M, Cienfuegos R (2020) History and features of trans-oceanic tsunamis and implications for paleo-tsunami studies. Earth Sci Rev 202:103112. https://doi.org/10.1016/j.earscirev.2020.103112
32. Sheehan AF, Gusman AR, Satake K (2019) Improving forecast accuracy with tsunami data assimilation: the 2009 dusky sound New Zealand tsunami. J Geophys Res: Solid Earth 124:566–577. https://doi.org/10.1029/2018JB016575
33. Synolakis CE, Skjelbreia JE (1993) Evolution of maximum amplitude of solitary waves on plane beaches. J Waterw Port Coast Ocean Eng 119(3):323–342. https://doi.org/10.1061/(ASCE)0733-950X(1993)119:3(323)
34. Tanioka Y (2018) Tsunami simulation method assimilating ocean bottom pressure data near a tsunami source region. Pure Appl Geophys 175:721–729. https://doi.org/10.1007/s00024-017-1697-5
35. Wang D, Becker NC, Walsh D et al (2012) Real-time forecasting of the April 11, 2012 Sumatra tsunami. Geophys Res Lett 39(19):L19601. https://doi.org/10.1029/2012GL053081
36. Wang Y, Maeda T, Satake K, Heidarzadeh M, Su H, Sheehan AF, Gusman AR (2019) Tsunami data assimilation without a dense observation network. Geophys Res Lett 46(4):2045–2053. https://doi.org/10.1029/2018GL080930
37. Wang Y, Satake K, Cienfuegos R, Quiroz M, Navarrete P (2019) Far-field tsunami data assimilation for the 2015 Illapel earthquake. Geophys J Int 209:514–521. https://doi.org/10.1093/gji/ggz309
38. Williamson A, Newman A (2018) Limitations of the resolvability of finite-fault models using static land-based geodesy and open-ocean tsunami waveforms. J Geophys Res: Solid Earth 123:9033–9048. https://doi.org/10.1029/2018JB016091
39. Williamson A, Newman A, Cummins P (2017) Reconstruction of coseismic slip from the 2015 Illapel earthquake using combined geodetic and tsunami waveform data. J Geophys Res: Solid Earth 122:2119–2130. https://doi.org/10.1002/2016JB013883
40. Yamazaki Y, Cheung KF (2011) Shelf resonance and impact of near-field tsunami generated by the 2010 Chile earthquake. Geophys Res Lett 38:L12605. https://doi.org/10.1029/2011GL047508
41. Yang Y, Dunham EM, Barnier G, Almquistj M (2019) Tsunami wavefield reconstruction and forecasting using the ensemble Kalman Filter. Geophys Res Lett 46:853–860. https://doi.org/10.1029/2018GL080644
42. Yang Z, Sheehan AF, Collins JA, Laske G (2012) The character of seafloor ambient noise recorded offshore New Zealand: results from the MOANA ocean bottom seismic experiment. Geochem Geophys Geosyst 13:Q10011. https://doi.org/10.1029/2012GC004201

Chapter 4
Real-Time Tsunami Detection Based on Ensemble Empirical Mode Decomposition (EEMD)

4.1 EEMD

Time–frequency analysis can characterize and manipulate signals in time and frequency domains. The EEMD is an adaptive time–frequency analysis method that is widely adopted in the atmospheric and oceanic sciences, for investigating parameters such as the variation of sea level pressure index, irregular coastal water waves, and the spatial–temporal change in air pollutants [4, 13, 22]. The decomposition assumes that at any given time, the original data has many coexisting oscillatory modes of significantly different frequencies superimposed on each other.

The Empirical Mode Decomposition (EMD) is the predecessor of EEMD. It decomposes the time series into a finite number of basis, defined as Intrinsic Mode Functions (IMFs). IMFs represent the natural oscillatory modes embedded in the signal and work as the basis functions. The number of IMFs is decided by the length of the data [22]. The IMFs need to satisfy the following two conditions: (1) The number of extrema and zero crossings must either be equal or differ at most by one. (2) The mean value of the upper envelope and lower envelope should be zero [10, 11]. Further details of EMD can be found in [11]. Here, I briefly describe the decomposition process as the following steps (Fig. 4.1):

(1) Find the local maxima and minima of the original time series.
(2) Connect all the local maxima (minima) by a cubic spline line as the upper (lower) envelope.
(3) Subtract the mean value of two envelopes from the original time series.
(4) Iterate these processes until the extracted signals satisfy the two conditions of IMFs (i.e., if the two conditions are satisfied, the residue becomes the raw IMF Otherwise, it enters the next iteration and repeats Steps 1–3 again to the residue).

After obtaining the 1st raw IMF, I subtract it from the original input signal and repeat the iteration above to extract the 2nd raw IMF and 3rd raw IMF. The number of raw IMFs, n_r, depends on the length of data: $n_r = \log_2 N_L$, where N_L is the length of the input signal [22]. The raw IMFs are listed in order of descending frequency.

Y. Wang, *Tsunami Data Assimilation for Early Warning*, Springer Theses,
https://doi.org/10.1007/978-981-19-7339-0_4

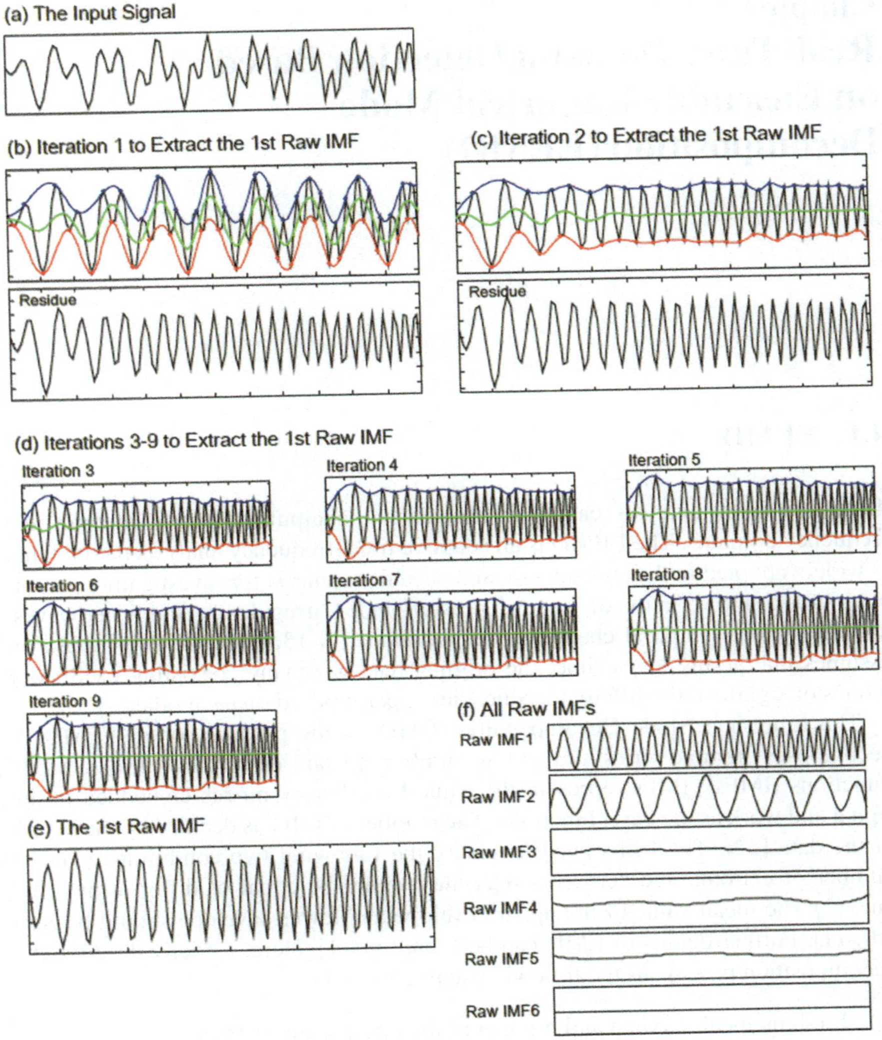

Fig. 4.1 Empirical Mode Decomposition of the input signals. The decomposition of an input signal (**a**) is iterated nine times until the residue satisfies the two conditions of Intrinsic Mode Functions (IMFs) (**b**–**e**). Blue and red dots represent local maxima and minima. The blue curve represents the upper envelope that connects all the local maxima, the red curve represents the lower envelope that connects all the local minima, and the green curve represents the mean value of two envelopes. Then it becomes the 1st raw IMF (**e**). After that, the 1st raw IMF is subtracted from the input signal and the iteration is repeated to extract other raw IMFs (**f**). Reprinted from Wang et al. [21] Copyright 2020

Short-period oscillation tends to appear in the lower mode IMFs, whereas higher-mode IMFs mostly contain long-period signals. The dominant period of each mode is not fixed and depends on the characteristics of the input signal.

To make it more robust to noise, Wu and Huang [22] modified the EMD by introducing the noise-assisted data analysis and proposed the EEMD. They used the mean of an ensemble of EMD trials, each consisting of the signal plus white noise of finite amplitude.

Compared with other time–frequency analysis methods like Fast Fourier transform or wavelet analysis, the most pronounced characteristic of EEMD is that the decomposition is determined by the characteristics of the original time series itself rather than a pre-determined basis (e.g., trigonometric functions). Therefore, it is an adaptive signal processing method that can be applied to nonlinear and non-stationary processes [12]. Moreover, unlike causal filters, EEMD does not cause any change of the tsunami peak by group delay. These advantages make it suitable for real-time tsunami detection.

To employ EEMD in a real-time tsunami detection algorithm, I follow several steps. The first step is the quality control of real-time data. I examine if there are any technical problems with the observational devices. Next, the time series in the past 3 h are selected: the segment from $y(t - 3h)$ to $y(t)$, where y is the value of tsunami height, and t is the time. I use the same segment length as NOAA's algorithm [15]. These processes can be done in real time. Then, I conduct EEMD to the 3 h segment and obtain the corresponding raw IMFs. Each of the raw IMF has different frequency characteristics. However, it is important to note that these raw IMFs are not orthogonal, and therefore some neighboring raw IMFs with similar frequency characteristics should be combined artificially [13, 22]. The input 3 h time series has 180 data points, and hence the decomposition of the 3 h time series generates seven raw IMFs, listed in descending order of frequency. Short-period oscillation tends to appear in the lower mode of IMFs, whereas the higher-mode IMFs mostly contain long-period signals (Fig. 4.2, left column). I plot their amplitude spectrum in the middle column to show the frequency characteristics. The 1st raw IMF has a dominant frequency of over 0.01 Hz (<100 s). The 2nd to 4th raw IMFs have dominant frequencies of 0.002–0.01 Hz (500–100 s). The 5th to 7th raw IMFs have dominant frequencies of approximately 0.001 Hz (1000 s). Hence, I use the 1st raw IMF as IMF1, combine the 2nd to 4th raw IMFs as IMF2, and combine the 5th to 7th raw IMFs as IMF3. Each IMF represents different components. The tsunami signal is presented in IMF2. The combination of raw IMFs is automatic and common for different events.

Finally, I compare the IMF amplitude at the last moment with a threshold value. If the Last-Moment IMF (LM-IMF) exceeds the threshold, the algorithm will declare the tsunami arrival. Otherwise, these steps are repeated for the next interval, with the new segment from $y(t - 3h + \Delta t)$ to $y(t + \Delta t)$, where Δt is the sampling interval. The IMFs are recalculated, and the amplitudes of LM-IMF are compared again, similar to the automatic phase detection method of seismograms, STA/LTA [1].

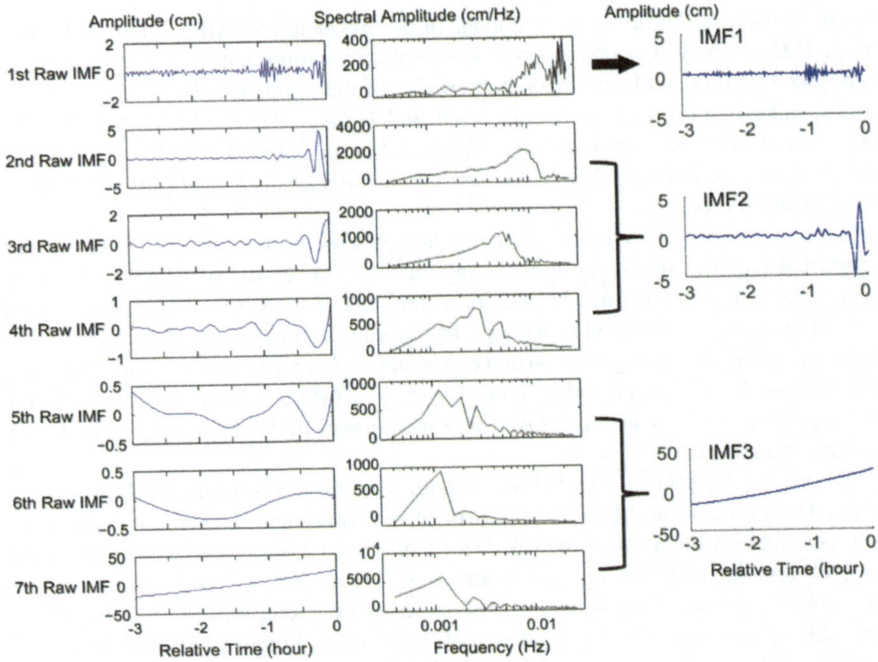

Fig. 4.2 Combination of raw Intrinsic Mode Functions (IMFs). The input signal is the sea-level record of station TM1 from 19:00 to 22:00 (UTC), November 21, 2016. It is decomposed into seven raw IMFs (left column). The frequency characteristics of raw IMFs are plotted in the middle column. I combine the raw IMFs with similar frequency characteristics artificially and obtain three IMFs (right column). Reprinted from Wang et al. [21] Copyright 2020

4.2 Validation Test—2016 Fukushima Earthquake

To validate the real-time tsunami detection algorithm based on EEMD, I use the data of the 2016 Fukushima earthquake (M 7.4) as an example. The earthquake occurred on November 21, 2016, at 20:59:47 UTC, offshore Fukushima Prefecture, Japan. Based on the JMA earthquake catalog, the hypocenter of the earthquake was at 37.355°N, 141.604°E, at a depth of 25 km ([6], Fig. 4.3a). The earthquake generated a moderate tsunami, with the maximum tsunami amplitude of 1.4 m measured at Sendai Port. It was the largest tsunami event since the 2011 Tohoku earthquake [18].

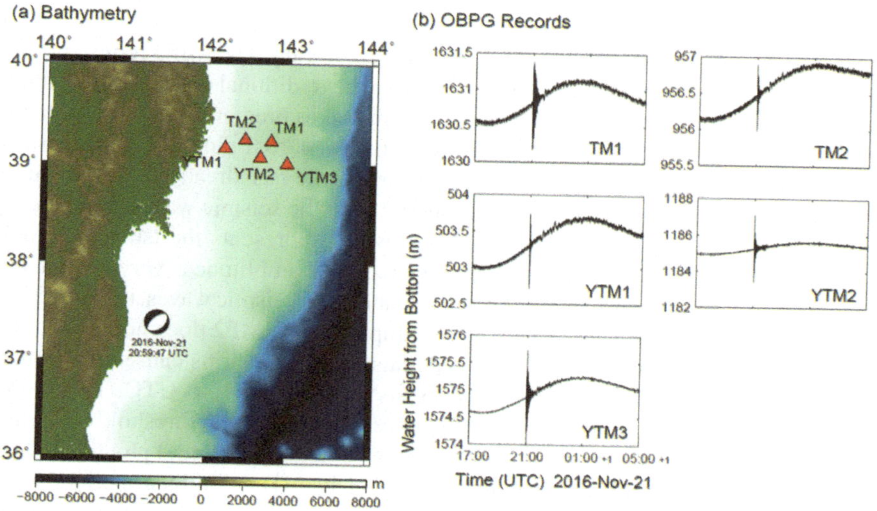

Fig. 4.3 Bathymetry and the Offshore Bottom Pressure Gauge (OBPG) records of the 2016 Fukushima earthquake. **a** Map of the epicenter of the 2016 Fukushima earthquake and OBPGs that recorded the tsunami. The locations of OBPGs are marked by red triangles. **b** The OBPG records of the event. These records have already been converted from water pressure to water height from the bottom. The tsunami signals are mixed up with tidal components and seismic waves. Reprinted from Wang et al. [21] Copyright 2020

4.2.1 Observed Tsunami Data

The tsunami is clearly recorded by OBPGs offshore Iwate Prefecture: TM1, TM2, YTM1, YTM2, and YTM3 (Fig. 4.3b). The sampling rates are 1 s for the two TM gauges and 0.1 s for the three YTM gauges.

For all five OBPGs, I convert the water pressure to water height by assuming hydrostatic pressure at the OBPG depth [6]. I resample the data by calculating the 1 min average for each OBPG. Then, I select a 3 h segment for EEMD. I test the algorithm as if it is processed in "real time" and assume no priori information about the earthquake event. I start the EEMD process approximately 1 h before the earthquake, with the first segment from 17:00 to 20:00 UTC. I obtain raw IMFs and combine the neighboring ones. Then, I compare the amplitude of LM-IMF2 with a threshold of 2.0 cm to decide whether the tsunami has arrived or not.

4.2.2 Results

In Fig. 4.4a, I plot the EEMD results of TM1 at 20:00 UTC, before the earthquake. At this moment, there is no seismic wave nor tsunami waves. The IMFs have a very small amplitude. IMF1 represents high-frequency components, including the background noise and the white noise artificially added for ensembling. The IMF3 represents the tide, oceanic, and meteorological signals. Unlike the traditional detection algorithm, the new algorithm does not need a theoretical model to predict tides. Instead, these signals are removed automatically by EEMD. In Fig. 4.4b, I plot the EEMD results of TM1 at 21:30 UTC. At this moment, the earthquake has already occurred, but it is before the tsunami arrival. The results suggest that the seismic waves are detected by OBPGs and represented in IMF1. IMF2, which represents the tsunami signal, still has a very small amplitude. Although the recorded amplitude is very large in the original time series because of ground motion caused by seismic waves, the algorithm does not confirm the tsunami arrival as the amplitude of IMF2 does not exceed the threshold. Therefore, it successfully avoids false tsunami alarms caused by ground motion. In Fig. 4.4c, I plot the EEMD results of TM1 at 21:46 UTC. This is the first moment when the amplitude of IMF2 exceeds the negative threshold (−2 cm). Hence, I determine that the tsunami arrives at this time. In Fig. 4.4d, I plot the EEMD results at 22:00 UTC, 14 min after the tsunami arrival. The dominant period bands of each IMF is adaptively determined by the characteristic of the input signal. It is different from digital filtering that has a specific period band. In this retroactive study, the dominant period of IMF1 is less than 100 s, the dominant period of IMF2 is 100–800 s, and the dominant period of IMF3 is above 1000 s.

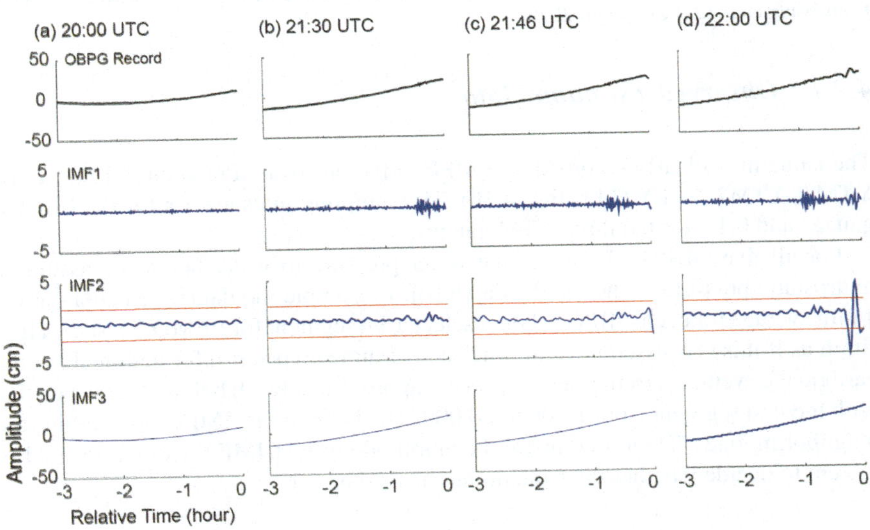

Fig. 4.4 Demonstration of the real-time tsunami detection algorithm. The Offshore Bottom Pressure Gauge records and Ensemble Empirical Mode Decomposition (EEMD) results of TM1 at 20:00 UTC (**a**), 21:30 (**b**), 21:46 (**c**), and 22:00 (**d**) are plotted, representing the moment before an earthquake, after an earthquake, and at and after tsunami arrival, respectively. Intrinsic Mode Function 2 (IMF2) represents the tsunami signal. In real-time operation, the EEMD results are updated continuously. When the amplitude of Last-Moment IMF2 exceeds the threshold, tsunami arrival is confirmed. Reprinted from Wang et al. [21] Copyright 2020

4.2.3 Evaluation of Detection: Comparison with Post-Processed Waveforms

I also compare the real-time detected results with post-processed waveforms. The series of real-time detected results of EEMD are LM-IMF2 at different times. They are used to determine the tsunami arrival and characterize the tsunami amplitude as if in "real time". For the post-processed waveforms, I use polynomial fitting to remove the tidal components, following [6, 9]. Additionally, I use a band-pass filter to process the time series, with the corner frequencies at 0.000667 and 0.0083 Hz (1500 and 120 s). The post-processed waveforms can characterize the tsunami arrival time and amplitude accurately, but they cannot be applied in real time. In TM1, the post-processed waveform shows that the tsunami arrives at 21:45 UTC, which is slightly earlier than the real-time detected results (<1 min; Fig. 4.5). Hence, the EEMD algorithm could detect the tsunami arrival at TM1 in real time, with a very short detection delay. For comparison, I also use a casual filter within the same frequency band to process the time series as if in "real time". Though the causal filters can be applied in real time, the detection delay (~3 min) is larger than that of EEMD and the tsunami amplitude is underestimated.

I also plot the detected results at the other four OBPGs and compare them with the post-processed waveforms. For most OBPGs, there is only a very short detection delay. The average tsunami detection delay is less than 1 min, whereas the tsunami wave period is approximately 10 min in this event. Compared with the average detection delay of one wave period of the NOAA's algorithm [3, 15], the algorithm can

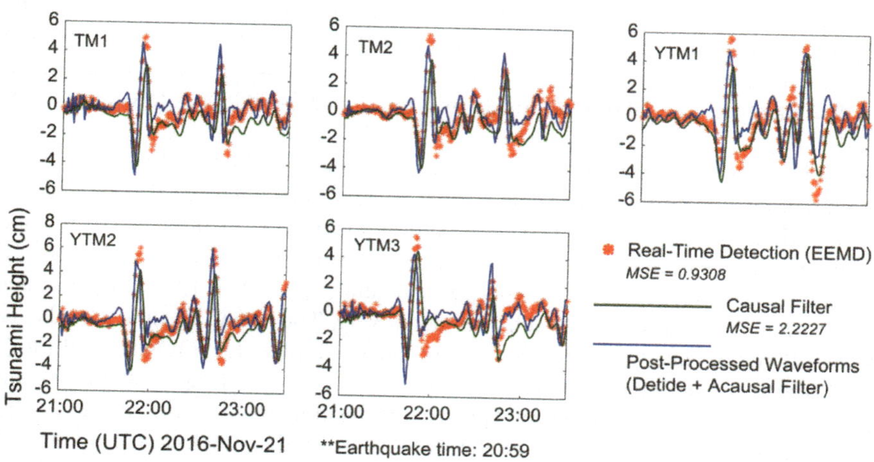

Fig. 4.5 Comparison of the real-time detected results and post-processed waveforms. The red stars represent the amplitude of Last-Moment Intrinsic Mode Function 2, calculated by Ensemble Empirical Mode Decomposition. The dark green curves present the tsunami waveforms that are processed with a causal filter as if operated in "real time". The blue curves represent the tsunami waveforms that are post-processed with an acausal filter

detect the tsunami arrival with a shorter delay. As for tsunami amplitude characterization, the real-time detected results of EEMD slightly underestimate the amplitude of the negative phase of the first tsunami peak, but they slightly overestimate the amplitude of the positive phase. Generally, they match quite well with the post-processed waveforms, with a mean-square error of 0.9308. Whereas, the real-time detected results of causal filter are less consistent with the post-processed waveforms, with a mean-square error of 2.2227. The amplitude of the positive phase is underestimated and there exist evident discrepancies in the following peaks.

4.3 Discussion

4.3.1 Applications to Extreme Cases

To test the applicable range of the real-time tsunami detection algorithm, I apply it to extreme tsunami cases. First, I adopt the 2011 Tohoku earthquake (M 9.0) as an example of extremely large tsunami. The earthquake occurred at 05:46:18 on March 11, 2011 (UTC), at 38.103°N, 142.860°E, at a depth of 24 km, according to JMA. At the time of this extremely large event, only stations TM1 and TM2 were installed offshore of the Tohoku region. The two OBPGs recorded the first peak of the tsunami (Fig. 4.6). However, at approximately 30 min after the earthquake, the devastating tsunami damaged the transmission station on land, and the two OBPGs lost connection. In the short records, the pressure data of TM1 and TM2 include the effects of seismic waves, tsunami, ocean tides, the vertical deformation of the seafloor, and atmospheric and oceanographic disturbances [8, 19]. After removing the high-frequency components, both records show a two-stage tsunami, an initial gradual increase for ~10 min, followed by a rapid increase within a few minutes [5, 14, 17, 19].

Similar to the application to the 2016 Fukushima earthquake, I apply the algorithm to the records of the two OBPGs, as if in "real time." Figure 4.7 demonstrates the EEMD results at 06:16:00 UTC (i.e., approximately 30 min after the earthquake origin time). IMF1 represents the high-frequency components, including the seismic waves and the first part of the two-stage tsunami [5, 14, 19]. EEMD determines the dominant period bands of each IMF adaptively. The strong seismic signals are partly filtered out in advance because of the data resampling (averaging over 1 min). IMF2 represents the main tsunami signals. IMF3 is the low-frequency component that represents the tidal components and the sea level increase owing to the vertical deformation of the seafloor. Though the IMF2 is affected by the long-period ground motion for this extremely large earthquake, it still characterizes the tsunami arrival time and amplitude. Therefore, it can confirm the tsunami arrival and provide information for tsunami early warning of such a large event. Reprinted from Wang et al. [21] Copyright 2020.

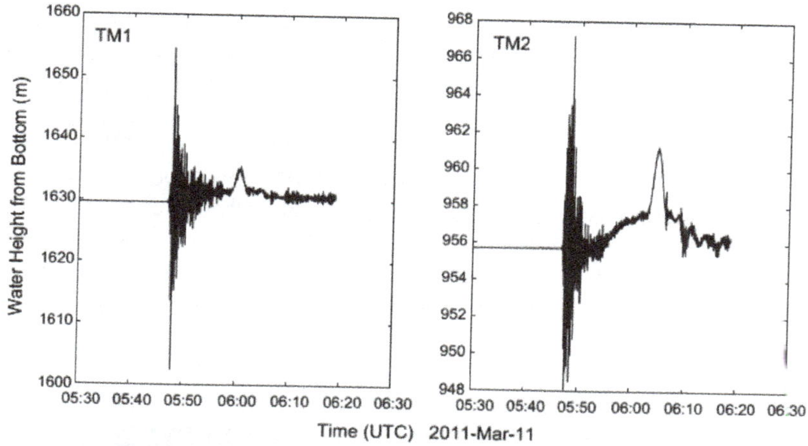

Fig. 4.6 Records of stations TM1 and TM2 during the 2011 Tohoku earthquake. Reprinted from Wang et al. [21] Copyright 2020

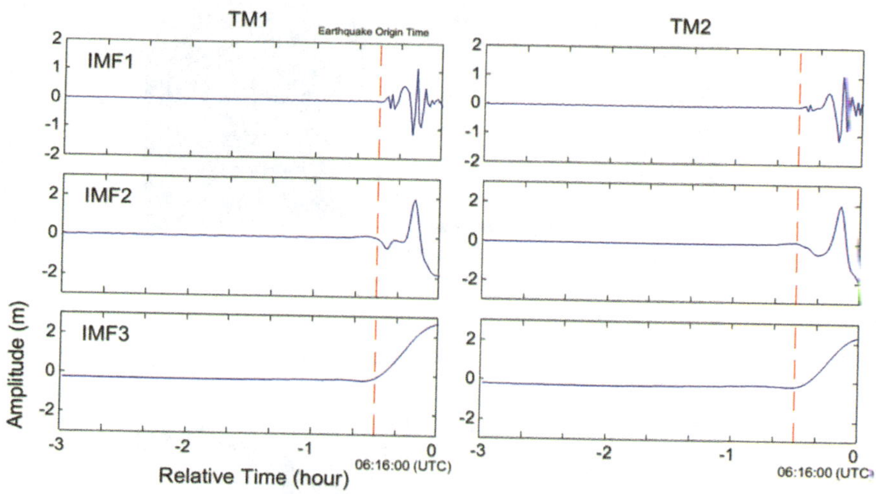

Fig. 4.7 Ensemble Empirical Mode Decomposition (EEMD) results of stations TM1 and TM2. The EEMD results are calculated at 06:16:00 (UTC), approximately 30 min after the 2011 Tohoku earthquake origin time. Reprinted from Wang et al. [21] Copyright 2020

The algorithm is also tested for a micro tsunami event by considering the 1998 off Sanriku earthquake (M 6.4). The earthquake occurred off the Sanriku region at 18:18:09 on May 30, 1998 (UTC) at 39.028°N, 143.847°E, at a depth of 11 km according to JMA (Fig. 4.8a). A micro tsunami was generated and observed by OBPGs TM1 and TM2 (Fig. 4.8b). Hino et al. [9] conducted a retroactive study and extracted the tsunami signals from the records of the OBPGs. The maximum peak-to-peak amplitude observed is approximately 1.5 cm [9], and it is in good agreement with the numerical simulation. I apply the real-time tsunami detection algorithm to the OBPG records and extract tsunami signals from the original time series.

Figure 4.9 demonstrates the EEMD results at 21:00:00 UTC. The IMF1 represents the high-frequency components. Seismic waves could be observed after the earthquake, but the strong seismic signals are partly filtered out because of the 1 min data resampling. IMF3 is the low-frequency component, which shows the tidal signals. IMF2 is supposed to represent the tsunami signals. However, the tsunami cannot be observed in IMF2. The absolute value of the IMF amplitude never exceeds 0.01 m.

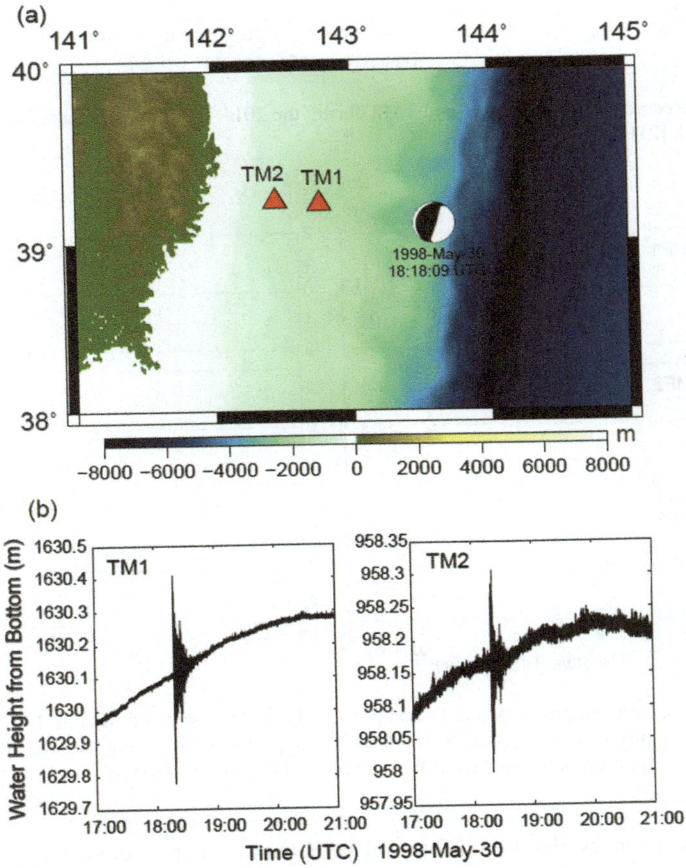

Fig. 4.8 Bathymetry and the Offshore Bottom Pressure Gauge (OBPG) records of the 1998 off Sanriku earthquake. **a** Map of the epicenter of the 1998 off Sanriku earthquake and OBPGs that recorded the tsunami. The locations of OBPGs are marked by red triangles. **b** The records of stations TM1 and TM2 during the 1998 off Sanriku earthquake. Reprinted from Wang et al. [21] Copyright 2020

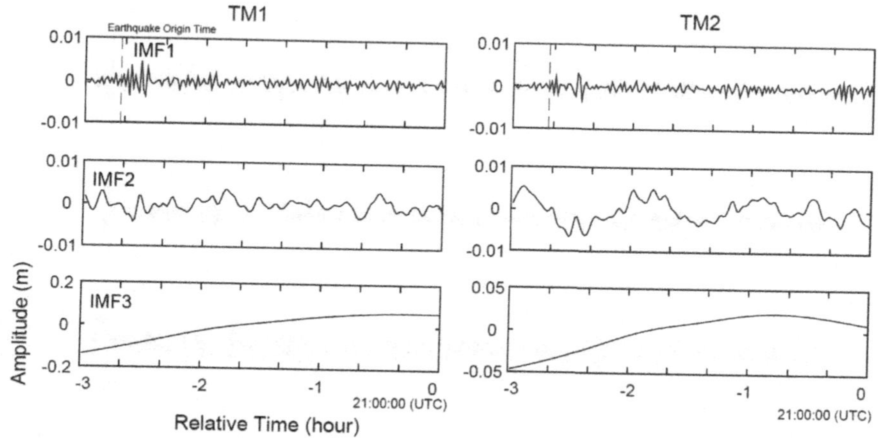

Fig. 4.9 Ensemble Empirical Mode Decomposition (EEMD) results of stations TM1 and TM2. The EEMD results are calculated at 21:00:00 (UTC), around 2 h 30 min after the 1998 off Sanriku earthquake. Reprinted from Wang et al. [21] Copyright 2020

It thus does not reach the threshold of 2 cm during the event, so I could not detect the tsunami arrival by the algorithm. If I set a smaller threshold, there will be too many false alarms to give reliable tsunami detection. Therefore, for such small events with a magnitude of approximately M 6.4, which are unlikely to cause tsunami damage, the algorithm does not work for real-time tsunami detection.

4.3.2 *False Alarms and Missed Alarms*

The EEMD algorithm performance is estimated by evaluating the detection probability, detection delay, and occurrence rate of false alarms [3]. It is very important to choose an appropriate threshold for the tsunami detection algorithm. In the application to the 2016 Fukushima earthquake, I use a threshold of 2.0 cm, which is slightly smaller than the threshold of NOAA's method (3.0 cm, [15]). A larger threshold decreases the occurrence rate of false alarms, but it also makes the algorithm less sensitive to moderate tsunamis.

To find a proper threshold of the real-time tsunami detection algorithm, I calculate the LM-IMF2 for one month before the 2016 Fukushima earthquake. I use the observational records of the same five OBPGs. Starting from the data of October 22, 2016, I repeat the EEMD process as if in "real-time operation" and calculate the LM-IMF2 of each trial (Fig. 4.10).

During the one-month period before the earthquake, the value of LM-IMF2 has a very small variation, within the range of −3 to 3 cm. When the tsunami arrives after the earthquake, the value of LM-IMF2 immediately increases, and the algorithm

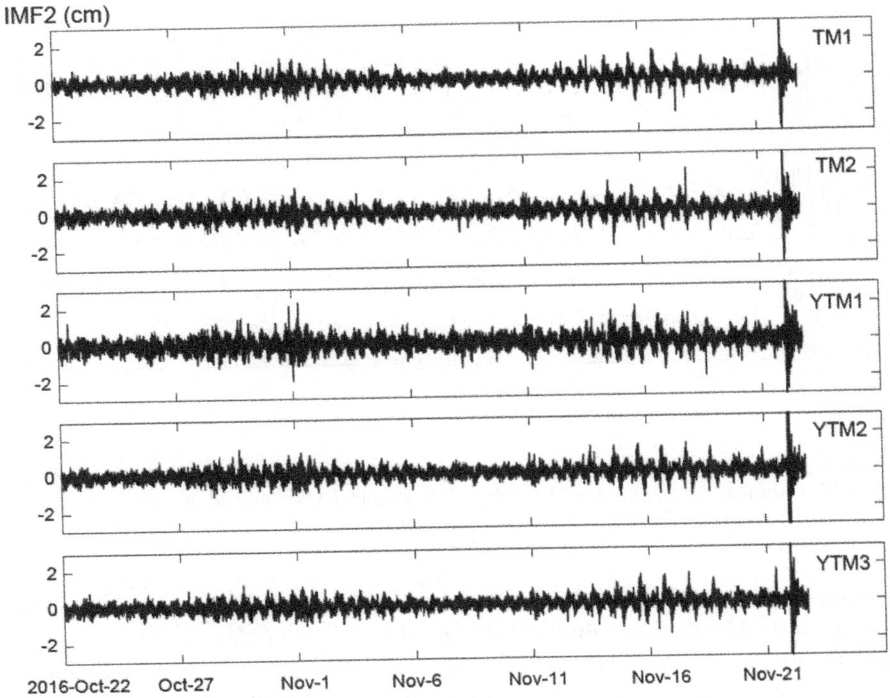

Fig. 4.10 Last-Moment Intrinsic Mode Function 2 of five Offshore Bottom Pressure Gauges for one month before the 2016 Fukushima earthquake. Reprinted from Wang et al. [21] Copyright 2020

detects the tsunami. As the LM-IMF2 never exceeds the absolute value of 3 cm in the one-month records, a threshold of 3 cm will never cause a false alarm in this period. This setting is similar to the threshold of NOAA's algorithm. Though there may be some missed alarms for very small tsunamis, it could effectively detect tsunamis that could result in potential flooding. On the other hand, if a stricter threshold is set (i.e., 2 cm in this study), the possibility of missed alarms will decrease, but false alarms will increase. For example, a threshold of 2 cm will lead to false alarms: three times at YTM1 and once at TM2. The false alarms of non-tsunami perturbation may result from oceanographic or meteorological phenomena like storm surges [2, 7]. If their frequency characteristics are similar to tsunamis, it is likely that these signals are also extracted in IMF2, leading to a false alarm. Therefore, it is important to select a threshold according to the requirements of the early warning system.

Furthermore, stations YTM1 and TM2 are the closest to the coast, and their depths are lower than those of the other three OBPGs (Fig. 4.3a). The amplitudes of tsunamis and other non-tsunami perturbations become larger according to Green's Law [16, 20]. As a result, stations with a smaller depth are more sensitive to perturbations and are more likely to give a false alarm. Therefore, it is also meaningful to consider different thresholds for stations at different depths.

References

1. Allen RV (1978) Automatic earthquake recognition and timing from single traces. Bull Seismol Soc Am 68(5):1521–1532
2. Bernard EN, Robinson AR (2009) Tsunamis, vol 15, 1st edn. Harvard University Press, Cambridge, p 28
3. Chierici F, Embriaco D, Pignagnoli L (2017) A new real-time tsunami detection algorithm. J Geophys Res: Oceans 122:636–652. https://doi.org/10.1002/2016JC012170
4. Dätig M, Schlurmann T (2004) Performance and limitations of the Hilbert-Huang transformation (HHT) with an application to irregular water waves. Ocean Eng 2004:1783–1834. https://doi.org/10.1016/j.oceaneng.2004.03.007
5. Fujii Y, Satake K, Sakai S, Shinohara M, Kanazawa T (2011) Tsunami source of the 2011 off the Pacific coast of Tohoku earthquake. Earth Planets Space 63:55. https://doi.org/10.5047/eps.2011.06.010
6. Gusman AR, Satake K, Shinohara M, Sakai S, Tanioka Y (2017) Fault slip distribution of the 2016 Fukushima earthquake estimated from tsunami waveforms. Pure Appl Geophys 174:2925–2943. https://doi.org/10.1007/s00024-017-1590-2
7. Hayashi Y (2008) Extracting the 2004 Indian Ocean tsunami signals from sea surface height data observed by satellite altimetry. J Geophys Res: Oceans 113:C01001. https://doi.org/10.1029/2007JC004177
8. Hayashi Y, Tsushima H, Hirata K, Kimura K, Maeda K (2011) Tsunami source area of the 2011 off the Pacific coast of Tohoku earthquake determined from tsunami arrival times at offshore observation stations. Earth Planets Space 63:809–813. https://doi.org/10.5047/eps.2011.06.042
9. Hino R, Tanioka Y, Kanazawa T et al (2001) Micro-tsunami from a local interpolate earthquake detected by cabled offshore tsunami observation in northeastern Japan. Geophys Res Lett 28:3533–3536. https://doi.org/10.1029/2001GL013297
10. Huang NE, Shen Z, Long SR (1999) A new view of nonlinear water waves: the Hilbert spectrum. Annu Rev Fluid Mech 31:417–457. https://doi.org/10.1146/annurev.fluid.31.1.417
11. Huang NE, Shen Z, Long SR, Wu MC et al (1998) The empirical mode decomposition and the Hilbert spectrum for nonlinear and non-stationary time series analysis. Proc R Soc A 454:903–995. https://doi.org/10.1098/rspa.1998.0193
12. Huang NE, Wu Z (2008) A review on Hilbert-Huang transform: method and its applications to geophysical studies. Rev Geophys 46:RG2006. https://doi.org/10.1029/2007RG000228
13. Liu M, Lin J, Wang Y, Sun Y et al (2018) Spatiotemporal variability of NO2 and PM2.5 over Eastern China: observational and model analyses with a novel statistical method. Atmos Chem Phys 18:12933–12952. https://doi.org/10.5194/acp-18-12933-2018
14. Maeda T, Furumura T, Sakai S, Shinohara M (2011) Significant tsunami observed at ocean-bottom pressure gauges during the 2011 off the Pacific coast of Tohoku Earthquake. Earth Planets Space 63:53. https://doi.org/10.5047/eps.2011.06.005
15. Mofjeld HO (1997) Tsunami detection algorithm. Not published paper. http://nctr.pmel.noaa.gov/tda_documentation.html
16. Satake K (2015) Treatise on geophysics, vol 4, 2nd edn. Elsevier, Oxford, pp 447–504
17. Satake K, Fujii Y, Harada T, Namegaya Y (2013) Time and space distribution of coseismic slip of the 2011 Tohoku earthquake as inferred from tsunami waveform data. Bull Seismol Soc Am 103:1473–1492. https://doi.org/10.1785/0120120122
18. Suppasri A, Leelawat N, Latcharote P, Roeber V et al (2017) The 2016 Fukushima earthquake and tsunami: local tsunami behavior and recommendations for tsunami disaster risk reduction. Int J Disaster Risk Reduct 21:323–330. https://doi.org/10.1016/j.ijdrr.2016.12.016
19. Tsushima H, Hirata K, Hayashi Y, Tanioka Y, Kimura K et al (2011) Near-field tsunami forecasting using offshore tsunami data from the 2011 off the Pacific coast of Tohoku earthquake. Earth Planets Space 63:821–826. https://doi.org/10.5047/eps.2011.06.052
20. Wang D, Becker NC, Walsh D et al (2012) Real-time forecasting of the April 11, 2012 Sumatra tsunami. Geophys Res Lett 39(19):L19601. https://doi.org/10.1029/2012GL053081

21. Wang Y, Satake K, Maeda T, Shinohara M, Sakai S (2020) A method of real-time tsunami detection using ensemble empirical mode decomposition. Seismol Res Lett 91:2851–2861. https://doi.org/10.1785/0220200115
22. Wu Z, Huang NE (2009) Ensemble empirical mode decomposition: a noise-assisted data analysis method. Adv Adapt Data Anal 1:1–41. https://doi.org/10.1142/S1793536909000047

Chapter 5
Real-Time Tsunami Data Assimilation of S-Net Pressure Gauge Records During the 2016 Fukushima Earthquake

5.1 Introduction

In previous chapters, I improve the tsunami data assimilation approach and propose a real-time tsunami detection algorithm using EEMD. In this chapter, I combine the real-time tsunami detection algorithm and data assimilation approach to forecast tsunamis on the coast.

The 2016 Fukushima earthquake is used as an example. In this event, the S-net pressure gauges recorded the sea-surface variations (Fig. 5.1). Although the observation systems suffered mechanical issues at a few OBPGs, the tsunami signals at many stations were extracted by Kubota et al. [4]. I apply the tsunami data assimilation approach to S-net pressure gauge records and forecast the tsunami waveforms at tide gauges in the Sanriku region. Although this study is still conducted retroactively, I use the real-time tsunami detection algorithm to process the raw records as if operating in real time. Then, I compare the forecasted waveforms with the real observations at tide gauges to validate the method.

5.2 Data and Methods

5.2.1 *Tsunami Records*

The raw S-net pressure gauge records are obtained from NIED. The raw data are water pressure records, with a sampling rate of 0.1 s (Fig. 5.2). I use the tsunami records from 28 stations: S2N01–12 and S3N11–26 (Figs. 5.1 and 5.2). Here, the prefixes S2 and S3 refer to the groups of S-net stations. At each station, the records of the two pressure gauges are almost identical [4]. I use the records of the second sensor (HP2WP) because there are more broken first sensors (HP1WP) than second sensors.

Y. Wang, *Tsunami Data Assimilation for Early Warning*, Springer Theses,
https://doi.org/10.1007/978-981-19-7339-0_5

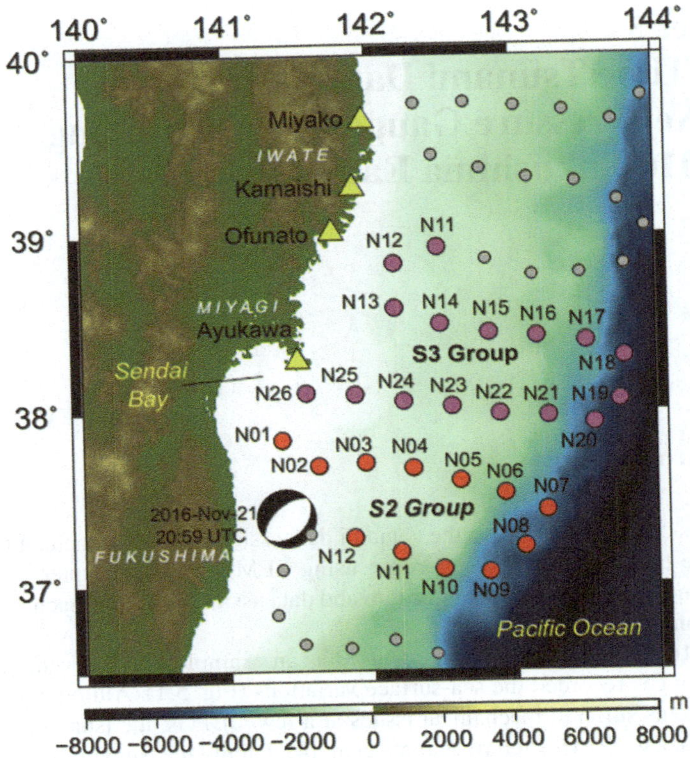

Fig. 5.1 Bathymetry map around the S-net pressure gauges and tide gauges. The red and pink circles indicate the Offshore Bottom Pressure Gauges (OBPGs) used for data assimilation, belonging to groups S2 and S3, respectively. The grey circles indicate other OBPGs that are not used in this study. Yellow triangles represent the tide gauges used for waveform comparison. The focal mechanism of the 2016 Fukushima earthquake is shown. Reprinted from Wang and Satake [9] Copyright 2021

The tide gauge records are used for waveform comparison to validate the method. I use the records of four tide gauges in the Sanriku region: Miyako, Kamaishi, Ofunato, and Ayukawa (Fig. 5.1). The tide gauge records of Miyako are digitized from a JMA report (https://www.jishin.go.jp/main/chousa/16dec/p16-e.htm). The records of Kamaishi are obtained from the Japan Oceanographic Data Center, with a sampling rate of 30 s. The records of Ofunato are obtained from IOC, with a sampling rate of 60 s. The digital records for Ayukawa are obtained from JMA, with a sampling rate of 60 s.

Although the S-net pressure gauges are also located to the south of the source region and recorded the 2016 tsunami waveforms, few coastal tide gauge stations to the south recorded the tsunami [2]. Therefore, I limit the assimilation area to the north of the source area, where a large number of observed tsunami waveforms are available, both offshore and on the coast.

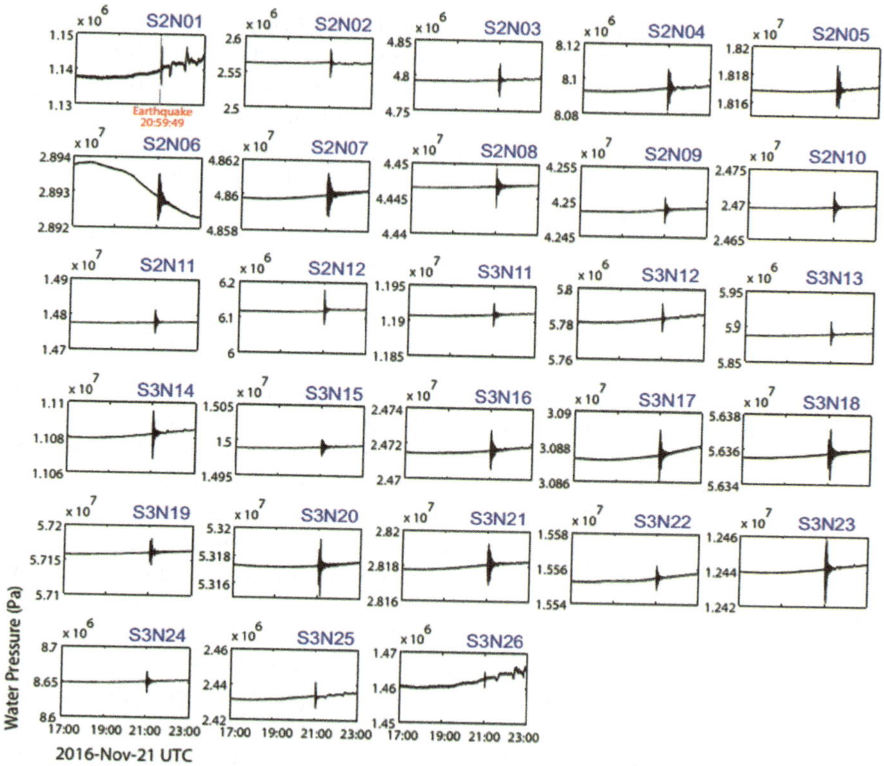

Fig. 5.2 Raw records from S-net pressure gauges during the 2016 Fukushima earthquake. Each station is equipped with two identical pressure gauge sensors to maintain redundancy. I use the records of the second sensor (HP2WP) in this study. Reprinted from Wang and Satake [9] Copyright 2021

5.2.2 Waveform Processing

The raw S-net records contain tidal components and seismic signals. Tsunami signals should be extracted before being used as an input for data assimilation. I compare two different methods of waveform processing: (1) the traditional post-processing method, including de-tiding and filtering [3], and (2) the real-time tsunami detection algorithm based on EEMD [10]. For both methods, I first convert the water pressure records to water height using the hydrostatic equation $p = \rho g h$, where p is the water pressure, and ρ is the density of seawater (1013 kg/m^3). For some short-wavelength tsunamis, the pressure on OBPGs may deviate from the hydrostatic relation when the wavelength of the sea-bottom deformation is not much greater than the sea depth [7].

The traditional post-processing method is widely adopted in retroactive studies to extract tsunami signals from OBPG sea-level records or tide gauges. First, the tidal components are removed by polynomial fitting. Then, a low-pass filter with a corner frequency of 0.01667 Hz (60 s) is applied to remove high-frequency components such as seismic waves [4]. However, as mentioned in Sect. 1.5, this method is not applicable in real-time operations because of the features of digital filters. On the contrary, EEMD can extract tsunami signals in real time. Over time, the decomposition is repeated continuously, and the time series of the LM-IMF2 can be used as the input for tsunami data assimilation, following the method in Sect. 4.1.

5.2.3 *Tsunami Data Assimilation*

I apply the tsunami data assimilation approach to forecast the waveforms at tide gauges based on offshore observations. In the study of the 2016 Fukushima earthquake, the assimilation begins at 21:00:30 UTC, when the tsunami is detected at the first OBPG (S2N12). In Table 5.1, I list the tsunami detection time at each S-net station determined by the algorithm. The waveforms at tide gauges are forecasted and updated continuously within the assimilation time window, at a time interval of 1 s. I test the time windows of 20 and 35 min for tsunami early warning and compare their performance.

To improve the assimilation speed, I adopt GFTDA, in which the Green's functions between the OBPGs and the tide gauges are computed in advance. Then, the forecasted waveforms are directly synthesized by the superposition of Green's functions. For the numerical simulation of tsunamis, the linear long-wave propagation model of *JAGURS* [1] is adopted. Nested grids with four layers are used for the numerical simulation. Layer 1 covers the entire region of the OBPGs and tide gauges, with a grid size of 18 arcsec (~555 m), and is derived from the bathymetry data of GEBCO_2014. Layers 2–4 are finer grids that cover the tide gauges, with a grid size of 6 arcsec (~185 m), 2 arcsec (~60 m), and 0.667 arcsec (~20 m), respectively. They are derived from the M7000 Digital Bathymetric Chart from the Japan Hydrographic

Table 5.1 Detected tsunami arrival time at each S-net station. The trigger time is in UTC. Reprinted from Wang and Satake [9] Copyright 2021

Station	Trigger time	Station	Trigger time	Station	Trigger time	Station	Trigger time
S2N01	21:01:00	S2N08	21:28:00	S3N13	21:45:30	S3N20	21:12:30
S2N02	21:05:00	S2N09	21:03:00	S3N14	21:41:30	S3N21	21:35:00
S2N03	21:01:00	S2N10	21:01:00	S3N15	21:35:30	S3N22	N/A
S2N04	21:01:30	S2N11	21:01:00	S3N16	21:40:00	S3N23	21:28:00
S2N05	21:22:30	S2N12	21:00:30	S3N17	21:05:30	S3N24	21:27:00
S2N06	21:29:00	S3N11	21:48:30	S3N18	21:10:00	S3N25	21:01:30
S2N07	21:30:30	S3N12	21:50:30	S3N19	21:05:00	S3N26	21:03:30

Association. The time step for calculating Green's functions is 0.25 s. The simulation is conducted on the EIC supercomputer at the Earthquake Research Institute, the University of Tokyo.

5.3 Results

5.3.1 Extracted Tsunami Signals

Figure 5.3 shows the tsunami signals extracted by the traditional post-processing method and real-time tsunami detection algorithm. In the traditional post-processing method, the S-net raw records are processed using polynomial fitting and low-pass filtering. In the real-time detection algorithm based on EEMD, the amplitude of LM-IMF2 represents the tsunami height characterized in real-time operations. For most OBPGs (e.g., S2N05, S3N11, and S3N26), the waveforms of the real-time tsunami detection algorithm match well with the waveforms of the traditional post-processing method in terms of both tsunami height and arrival time. These results are consistent with Sect. 4.2.3, where the raw records of five OBPGs owned by Earthquake Research Institute, the University of Tokyo, were processed using two similar methods. Hence, I confirm that the algorithm based on EEMD can detect the tsunami arrival and can accurately characterize the amplitude in real time.

In Table 5.1, I list the detected tsunami arrival time at each S-net station. The arrival time is determined by the real-time tsunami detection algorithm based on EEMD. After decomposition, I compare the value of last-moment IMF2 (LM-IMF2) with a threshold of 3.0 cm. The time that this threshold is exceeded is recorded as the detected tsunami trigger time. The data assimilation process begins after the first OBPG is triggered. As the OBPGs are near the source region, the tsunami is detected by OBPGs soon after the earthquake happens. At most stations (e.g., S3N11, S3N14), the detected arrival time is the moment when tsunami actually arrives, indicating that the real-time tsunami detection algorithm accurately detects the tsunami arrival.

However, at some stations close to the epicenter, there are some large drift components (e.g., S2N06) or abrupt steps (e.g., S2N11) at the earthquake origin time. According to the analysis of Kubota et al. [4], these drifts or steps are possibly caused by the mechanical issue of the observation system as the trend of the drifts or steps is completely identical in the pair of pressure sensors (i.e., HP1WP, HP2WP) at each S-net station. Similar steps were also found in the records of a much smaller tsunami (<1 cm) caused by the 2016 off Sanriku earthquake (Mw 6.0; [5]). The traditional post-processing method cannot completely remove these unexpected signals, and they also appear in the waveforms of the real-time detection algorithm at some stations (Fig. 5.3). The detected arrival time is affected by such unexpected signals at these stations, which may cause false tsunami alarms. Nevertheless, the real-time tsunami detection algorithm accurately characterizes the waveforms after the earthquake origin time, especially that of the first tsunami peak, which is very important

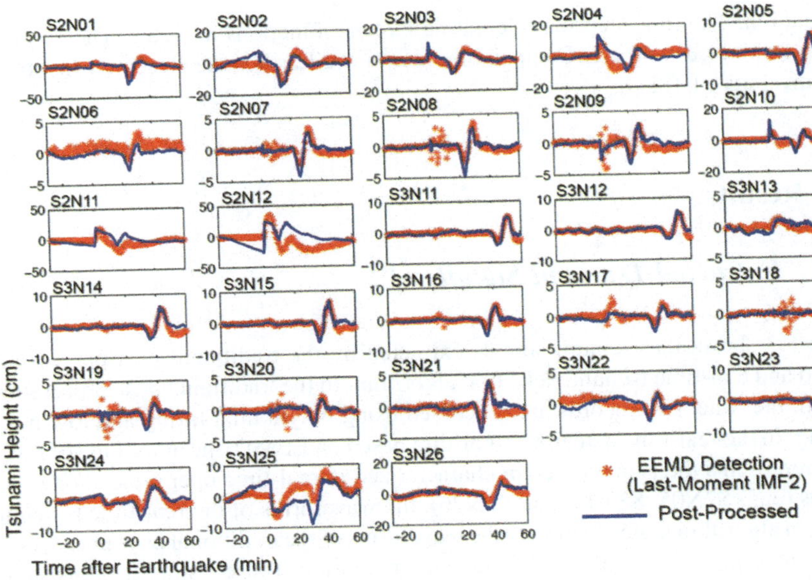

Fig. 5.3 Tsunami signals extracted by the traditional post-processing method (blue curves) and the real-time tsunami detection algorithm based on Ensemble Empirical Mode Decomposition (red stars). Reprinted from Wang and Satake [9] Copyright 2021

for data assimilation. In this study, the waveforms processed with the two methods are used separately as the input for tsunami data assimilation.

5.3.2 *Forecasted Tsunami Waveforms at Tide Gauges*

Figure 5.4 shows the comparison between the forecasted tsunami waveforms and the real observations. The waveforms are forecasted by tsunami data assimilation for time windows of 20 min (a) and 35 min (b). The actual tsunami arrives first at Ayukawa (57 min after the earthquake), and then at Ofunato (68 min), Kamaishi (77 min), and Miyako (80 min). At most tide gauges, the waveforms forecasted using the traditional post-processing method and real-time tsunami detection algorithm match well with real observations.

At Miyako, the data assimilation approach forecasts the arrival time of the first tsunami peak accurately. The observed first-peak amplitude is 26 cm, though the assimilation result with a 20 min time window may underestimate the first-peak amplitude. The 35 min time window has a better performance in forecasting the amplitude, especially for the results with input time series obtained using the real-time detection algorithm. Its forecasted first-peak amplitude is 18 cm. Moreover, the forecasted waveform is also consistent with the real observation in another peak of

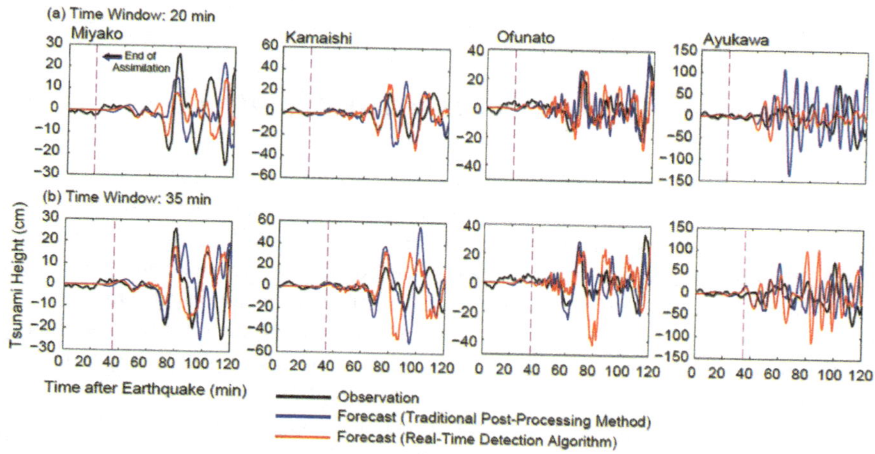

Fig. 5.4 Comparison of observed and forecasted tsunami waveforms. The assimilation time windows are 20 min (**a**) and 35 min (**b**). The black curves indicate the real observations. The blue curves represent the results forecasted by data assimilation with input time series obtained by the traditional post-processing method. The red curves represent the results forecasted by data assimilation with the input time series obtained using the real-time tsunami detection algorithm. Reprinted from Wang and Satake [9] Copyright 2021

the later phase (102 min). At Kamaishi, the forecasted waveforms generally match real observations, but the amplitude of the first tsunami peak is overestimated. This issue is discussed in Sect. 5.4.

At Ofunato, the forecasted waveforms match quite well with the real observations regarding the arrival time and amplitude. Both input time series lead to accurate forecasting by data assimilation. For example, for a 35 min time window, the observed amplitude of the first tsunami peak is 26 cm, and the forecasted amplitudes are 29 cm (by post-processing method) and 23 cm (by real-time detection algorithm). The forecast could be accurately made more than 30 min before the tsunami arrival. Nevertheless, the forecasted waveforms do not match well with the real observations at Ayukawa. The tsunami arrival time is not well predicted, and the forecasted results overestimate the amplitude of the first tsunami peak.

To evaluate the forecast accuracy quantitatively, I use Eq. (2.7). For a 20 min time window, the scores are 47% (by post-processing method) and 41% (by real-time detection algorithm) when considering all stations. For a 35 min time window, the scores are 60% and 74%, respectively. If I do not consider the Kamaishi and Ayukawa stations, the scores are 91% (by post-processing method) and 77% (by real-time detection algorithm) for a 20 min time window and they become 89% and 94%, respectively, for a 35 min time window.

5.4 Discussion

5.4.1 *Adoption of a Real-Time Detection Algorithm*

Compared with previous studies on tsunami data assimilation (e.g., [8, 11], the most notable feature of this work is the adoption of a real-time tsunami detection algorithm for waveform processing. I process the raw OBPG records as if operating in real time and use the processed waveforms as the input for data assimilation. The assimilation results with input obtained by the real-time tsunami detection algorithm perform well in tsunami forecasting. Hence, my experiment proves the feasibility of establishing a real-time tsunami early warning system based on the data assimilation approach. Although there are some unexpected signals at the earthquake origin time, as shown in Sect. 5.3.1, they do not significantly affect the performance of tsunami data assimilation as data assimilation is a self-correcting process. Over time, the effects of such unexpected signals are gradually corrected by assimilating more input data into the model, and the forecasted wavefield matches the real observations more accurately [6].

5.4.2 *Results of the Synthetic Experiment*

To further investigate the forecasting performance, I conduct a synthetic experiment using numerical simulations. Data assimilation for synthetic data can facilitate the separation of two factors of assimilation performance: imperfect modeling and insufficient observations. If there is a discrepancy between the observed and synthetic waveforms, and if the data assimilation performs better for synthetic data, the modeling is not perfect. On the other hand, if the observed and synthetic waveforms are similar, and the data assimilation does not perform well for synthetic data, then the offshore observations are insufficient for tsunami data assimilation.

I adopt the source model of [2], which is inverted from OBPG and tide gauge records. The bathymetry files are the same as those used to compute Green's functions. First, I run the tsunami propagation model *JAGURS* to compute the synthetic waveforms at S-net OBPGs and tide gauges. I compare the synthetic waveforms at four tide gauges with the real observations in Fig. 5.5. Next, I use the synthetic waveforms at the OBPGs as the input for tsunami data assimilation and forecast the waveforms at tide gauges. The forecasted waveforms and the synthetic results at tide gauges are compared in Fig. 5.6.

In Fig. 5.5, the synthetic first-peak amplitude at Kamaishi (42 cm) is much higher than the real observation (18 cm). However, the data assimilation with synthetic data works perfectly at Kamaishi (Fig. 5.6). This indicates that the discrepancies between the forecasted and observed waveforms at Kamaishi may have resulted from imperfect modeling because of bathymetry data and the source model. The Kamaishi tide gauge is located deep inside the harbor. Local bathymetry (e.g., breakwater and

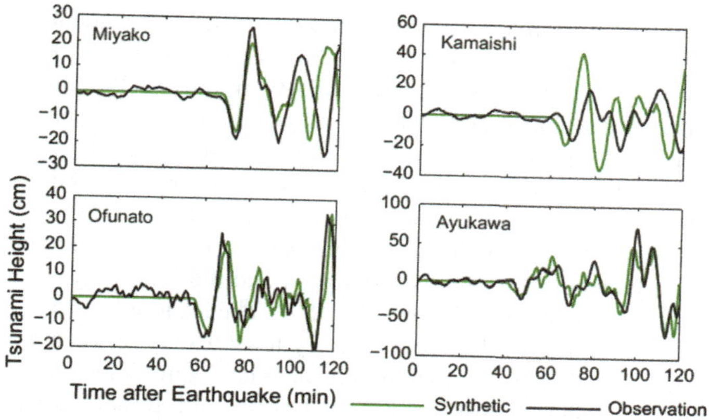

Fig. 5.5 Comparison of synthetic tsunami waveforms and real observations. The green curves represent synthetic tsunami waveforms, and the black curves represent real observations. The synthetic waveforms are generally consistent with the real observations at Miyako, Ofunato, and Ayukawa, but overestimate the tsunami amplitude at Kamaishi. Reprinted from Wang and Satake [9] Copyright 2021

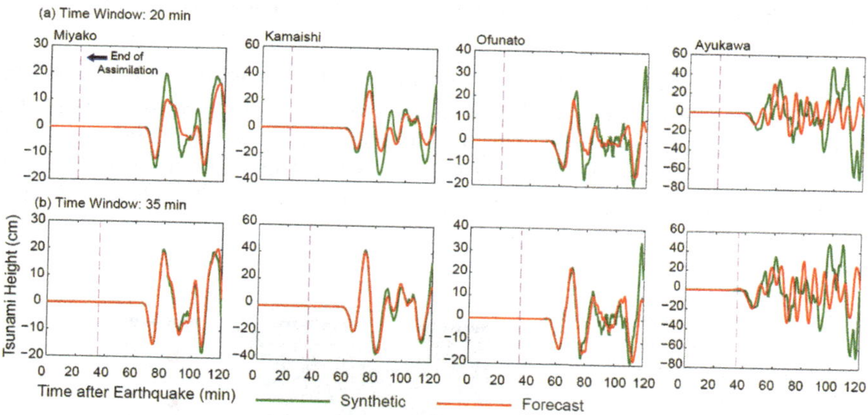

Fig. 5.6 Comparison of synthetic and forecasted tsunami waveforms. The green curves represent synthetic tsunami waveforms, and the red curves represent forecasted tsunami waveforms. The assimilation time windows are 20 min (**a**) and 35 min (**b**). The synthetic waveforms at tide gauges are used as the "observations" for comparison. Reprinted from Wang and Satake [9] Copyright 2021

wave attenuator) has a substantial effect on tsunami simulation, difficult to model even with the 0.667 arcsec grid (~20 m). Green's functions between the OBPGs and the Kamaishi tide gauge are also computed using these grid files; hence, the forecasted waveform synthesized by Green's functions overestimates the amplitude systematically. For a 20 min time window, there are fewer offshore data used for assimilation. This counteracts such systematic errors. For a 35 min time window, there are sufficient offshore data to make the systematic error evident. If we could obtain finer grids that represent the detailed bathymetry around the Kamaishi tide gauge, the accuracy of data assimilation could be improved. Furthermore, in my study, a Gaussian-distributed covariance is assumed for the correlation functions of both observation and computational errors when calculating the weight matrix $\boldsymbol{W}$ (Eq. 1.28). If the parameters of correlation functions are adjusted according to different data quality, the accuracy may also be improved.

However, the synthetic experiment still has unsatisfactory assimilation results at the Ayukawa tide gauge. The first tsunami peak and the following waveforms are not well predicted at both time windows (Fig. 5.6). The location of Ayukawa is quite close to the epicenter (Fig. 5.1). There are very few OBPGs between the epicenter and the tide gauge. Although data assimilation does not require source information, it requires sufficient offshore data to reconstruct the tsunami wavefield and facilitate the forecasting. Hence, the discrepancies at Ayukawa are caused by insufficient offshore observations. The installation of some OBPGs inside Sendai Bay is necessary for improving the forecasting ability.

5.4.3 Data Assimilation at Ayukawa with Assumed OBPGs

To prove the effectiveness of additional OBPGs, I add two assumed OBPGs inside Sendai Bay and use data assimilation to forecast the tsunami waveforms at the Ayukawa tide gauge (Fig. 5.7). The locations of two assumed OBPGs are V01 (38.20°N, 141.25°E) and V02 (38.00°N, 141.25°E). The synthetic waveforms at the OBPGs are used as the input for tsunami data assimilation. The other conditions are the same as those in the synthetic experiment.

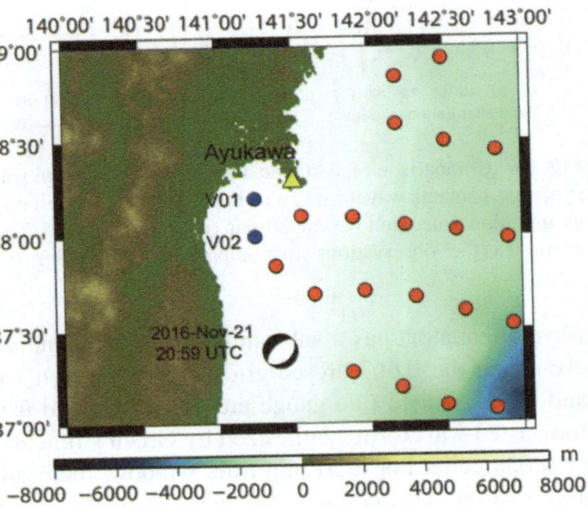

Fig. 5.7 Bathymetry map around Ayukawa tide gauge. The red circles indicate the real S-net Offshore Bottom Pressure Gauges (OBPGs) used for data assimilation. The blue circles indicate two hypothetical OBPGs: V01 and V02. The yellow triangle represents the Ayukawa tide gauge. The focal mechanism of the 2016 Fukushima earthquake is shown. Reprinted from Wang and Satake [9] Copyright 2021

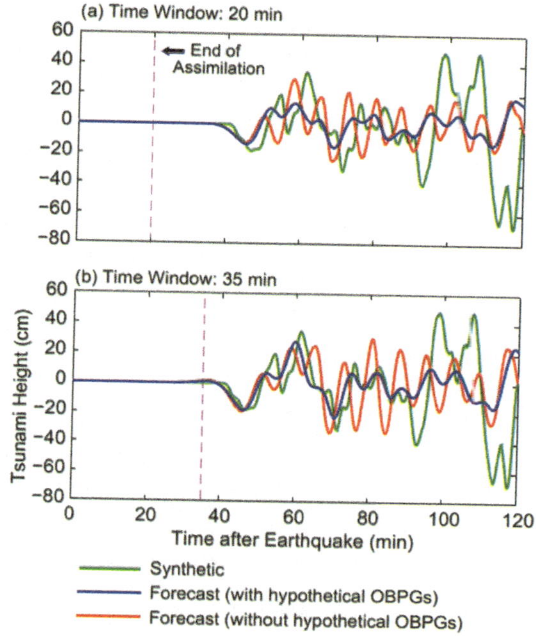

Fig. 5.8 Comparison between the synthetic and forecasted tsunami waveforms. The green curves represent synthetic tsunami waveforms, the blue curves represent forecasted tsunami waveforms with hypothetical Offshore Bottom Pressure Gauges (OBPGs), and the red curves represent forecasted tsunami waveforms without hypothetical OBPGs at the Ayukawa tide gauge. The assimilation time windows are 20 min (**a**) and 35 min (**b**). The synthetic waveform at Ayukawa is used as the "observation" for comparison. Reprinted from Wang and Satake [9] Copyright 2021

Figure 5.8 shows the comparison between the synthetic and forecasted tsunami waveforms. Here, the synthetic waveforms at tide gauges are used as the "observations" for comparison. I compare the forecasted results with and without two hypothetical OBPGs, for assimilation time windows of 20 and 35 min. For a 20 min assimilation time window, additional OBPGs inside Sendai Bay do not significantly improve the forecast accuracy, though the prediction of the first tsunami peak is better. However, for a 35 min assimilation window, the results with two hypothetical OBPGs have better performance. The first tsunami peak can be predicted better, and it works particularly well for the following waveform at 40–90 min after the earthquake. Hence, it is confirmed that installing OBPGs inside Sendai Bay is necessary to improve forecasting.

References

1. Baba T, Takahashi N, Kaneda Y, Ando K, Matsuoka D, Kato T (2015) Parallel implementation of dispersive tsunami wave modeling with a nesting algorithm for the 2011 Tohoku Tsunami. Pure Appl Geophys 172:3433–3472. https://doi.org/10.1007/s00024-015-1049-2
2. Gusman AR, Satake K, Shinohara M, Sakai S, Tanioka Y (2017) Fault slip distribution of the 2016 Fukushima earthquake estimated from tsunami waveforms. Pure Appl Geophys 174:2925–2943. https://doi.org/10.1007/s00024-017-1590-2

3. Heidarzadeh M, Harada T, Satake K, Ishibe T, Gusman AR (2016) Comparative study of two tsunamigenic earthquakes in the Solomon Islands: 2015 Mw 7.0 normal-fault and 2013 Santa Cruz Mw 8.0 megathrust earthquakes. Geophys Res Lett 43(9):4340–4349. https://doi.org/10.1002/2016GL068601
4. Kubota T, Chikasada NY, Tsushima H, Suzuki W, Nakamura T, Kubo H (2020) Tsunami analysis using the S-net pressure gauge records during the Mw 7.0 Off-Fukushima earthquake on 22 November 2016 to reduce the effects of tsunami-irrelevant pressure components. In: JpGU-AGU Joint Meeting 2020, HDS08-11
5. Kubota T, Saito T, Suzuki W (2020) Millimeter-scale tsunami detected by a wide and dense observation array in the deep ocean: fault modeling of an Mw 6.0 interplate earthquake off Sanriku, NE Japan. Geophys Res Lett 47:2. https://doi.org/10.1029/2019GL085842
6. Maeda T, Obara K, Shinohara M, Kanazawa T, Uehira K (2015) Successive estimation of a tsunami wavefield without earthquake source data: a data assimilation approach toward real-time tsunami forecasting. Geophys Res Lett 42:7923–7932. https://doi.org/10.1002/2015GL065588
7. Saito T, Kubota T (2020) Tsunami modeling for the deep sea and inside focal areas. Annu Rev Earth Planet Sci 48:121–145. https://doi.org/10.1146/annurev-earth-071719-054845
8. Sheehan AF, Gusman AR, Satake K (2019) Improving forecast accuracy with tsunami data assimilation: the 2009 dusky sound New Zealand tsunami. J Geophys Res: Solid Earth 124:566–577. https://doi.org/10.1029/2018JB016575
9. Wang Y, Satake K (2021) Real-time tsunami data assimilation of S-net pressure gauge records during the 2016 Fukushima earthquake. Seismol Res Lett 92:2145–2155. https://doi.org/10.1785/0220200447
10. Wang Y, Satake K, Maeda T, Shinohara M, Sakai S (2020) A method of real-time tsunami detection using ensemble empirical mode decomposition. Seismol Res Lett 91:2851–2861. https://doi.org/10.1785/0220200115
11. Wang Y, Satake K, Sandabata O, Maeda T, Su H (2019) Tsunami data assimilation of cabled ocean bottom pressure records for the 2015 Torishima volcanic tsunami earthquake. J Geophys Res: Solid Earth 124:10413–10422. https://doi.org/10.1029/2019JB018056

Chapter 6
Tsunami Early Warning System Using Data Assimilation of Offshore Data

6.1 Practical Implementation

In Chap. 5, the retroactive study of the 2016 Fukushima earthquake confirmed the applicability of real-time tsunami forecasting independent of the source. Therefore, a tsunami early warning system can be designed for Japan based on data assimilation.

The main target regions are the Tohoku and Nankai regions. These two regions face potential tsunami hazards, and there are dense OBPG networks with real-time data transmission: DONET and S-net. These networks represent reliable hardware for tsunami early warning. In general, four steps need to be taken to establish the tsunami early warning system using data assimilation of offshore data.

First, Green's functions between OBPGs and the coastal points (PoIs) are calculated and stored in advance. GFTDA is adopted to improve the efficiency of the system. Although the calculation of Green's functions is time-consuming, it does not affect the speed of the assimilation process.

Next, the real-time tsunami detection algorithm based on EEMD is used. It can detect the tsunami arrival without a priori earthquake information, and is thus suitable for tsunami data assimilation. The algorithm is performed continuously, and the LM-IMF2 is compared with the pre-defined threshold at each moment. Before the earthquake (or landslide or volcano eruption), the LM-IMF2 should remain below the threshold, and the system is in the *General Mode* (Fig. 6.1). The assimilation process does not begin. After an earthquake (or landslide or volcano eruption) event, as long as the tsunami does not arrive at OBPGs, the system remains in the General Mode.

When a tsunami arrives at OBPGs, the LM-IMF2 calculated by the EEMD algorithm will exceed the threshold. The early warning system is triggered and enters the *Assimilation Mode*. The pre-calculated Green's functions are superposed by matrix manipulation to predict the waveforms at the coast (Fig. 6.1). As time passes, more offshore tsunami data is observed and assimilated. The forecasted waveforms are updated continuously. Based on the forecasted results, we can decide whether to issue a tsunami warning or tsunami advisory to specific regions.

Y. Wang, *Tsunami Data Assimilation for Early Warning*, Springer Theses,
https://doi.org/10.1007/978-981-19-7339-0_6

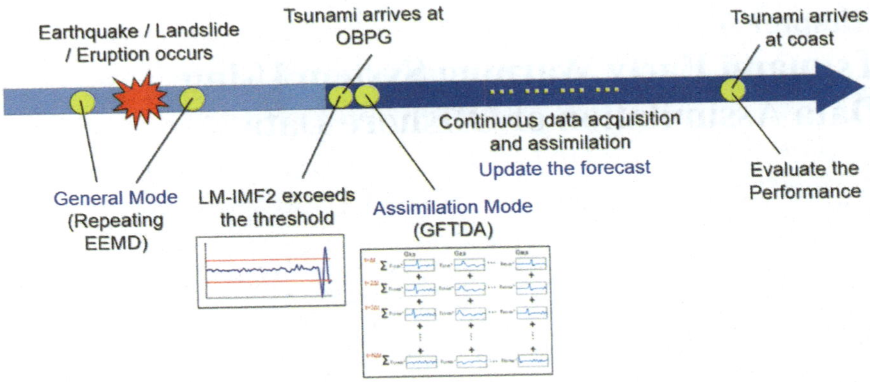

Fig. 6.1 Flow chart for the proposed tsunami early warning system based on the data assimilation approach

Finally, after the tsunami arrives at the coast, the performance of the tsunami early warning system is evaluated by comparing the forecasted results with real observations.

To summarize, the tsunami early warning system based on the data assimilation approach can provide early warning for coastal regions, regardless of the tsunami source. However, it is important to note that the goal of such a system is not to replace the traditional tsunami early warning system based on seismic observations [2]. Instead, it can be an important supplement to traditional early warning systems. If a tsunamigenic earthquake occurs, the traditional method can take advantage of the rapid availability of seismic data and make an initial warning quickly. When OBPGs detect the tsunami, the new system can start the data assimilation process and update the tsunami forecast by assimilating the offshore observations continuously. Additionally, it can also help with forecasting methods based on tsunami waveform inversion, such as the tFISH method operated by JMA [6]. Using the EEMD algorithm, the tsunami waveforms for inversion could be extracted directly without the need of a theoretical tide model, and the tFISH method can become more efficient.

Although this system is designed for Japan, it can be used as a reference in other tsunamigenic zones, especially in regions with tsunami events caused by non-earthquake origins like landslides or volcanoes. For example, the collapse of the Anak Krakatau Volcano in December 2018 generated a tsunami in the Sunda Strait, Indonesia, with a maximum run-up height of 13.5 m [5]. And the September 2018 Palu tsunami is widely believed to be of a dual landslide–earthquake source [1, 3]. Traditional tsunami early warning systems have difficulties in forecasting such non-earthquake tsunamis. To the contrary, the early warning system based on data assimilation does not require the construction of seismic/tsunami source model, nor even the detection of ground motion. Thus, it is very effective in non-seismic tsunamis. Additionally, the method of tsunami data assimilation with virtual stations can reduce the engineering cost of installing OBPGs in these regions.

6.2 Future Improvements

In the future, the tsunami early warning system based on data assimilation can be further improved by adopting deep learning techniques.

As mentioned in Chap. 4, the EEMD algorithm can detect the tsunami signals by comparing the LM-IMF2 with a pre-defined threshold. However, the threshold is set artificially according to experience. A higher threshold can cause more missed alarms, and a lower threshold increases the rate of false alarms. It is challenging to select an appropriate threshold. Therefore, we could use the deep learning technique to improve the real-time tsunami detection algorithm and to avoid setting a threshold artificially. Deep learning greatly improves the development of neural networks and has been widely used in earth science [4, 7]. After being trained, the deep learning model can detect the tsunami arrival and characterize the amplitude automatically instead of artificially setting a threshold for comparison. With the help of deep learning, the input for data assimilation can be more accurate, which would lead to better forecasts of the tsunami early warning system.

References

1. Carvajal M, Araya-Cornejo C, Sepúlveda I et al (2019) Nearly instantaneous tsunamis following the Mw 7.5 2018 Palu earthquake. Geophys Res Lett 46:5117–5126. https://doi.org/10.1029/2019GL082578
2. Duputel Z, Rivera L, Kanamori H et al (2011) Real-time W phase inversion during the 2011 off the Pacific coast of Tohoku earthquake. Earth Planets Space 63:5. https://doi.org/10.5047/eps.2011.05.032
3. Heidarzadeh M, Muhari A, Wijanarto AB (2018) Insights on the source of the 28 September 2018 Sulawesi tsunami, Indonesia based on spectral analyses and numerical simulations. Pure Appl Geophys 176:25–43. https://doi.org/10.1007/s00024-018-2065-9
4. Kuyuk HS, Susumu O (2018) Real-time classification of earthquake using deep learning. Procedia Comput Sci 140:298–305. https://doi.org/10.1016/j.procs.2018.10.316
5. Muhari A, Heidarzadeh M, Susmoro H et al (2019) The December 2018 Anak Krakatau Volcano Tsunami as inferred from post-tsunami field surveys and spectral analysis. Pure Appl Geophys 176:5219–5233. https://doi.org/10.1007/s00024-019-02358-2
6. Tsushima H, Hirata K, Hayashi Y, Tanioka Y, Kimura K et al (2011) Near-field tsunami forecasting using offshore tsunami data from the 2011 off the Pacific coast of Tohoku earthquake. Earth Planets Space 63:821–826. https://doi.org/10.5047/eps.2011.06.052
7. Zhang X, Zhang Q, Zhang G, Nie Z et al (2018) A novel hybrid data-driven model for daily land surface temperature forecasting using long short-term memory neural network based on ensemble empirical mode decomposition. Int J Environ Res Public Health 15:1032. https://doi.org/10.3390/ijerph15051032

Chapter 7
Summary

In this thesis, I improve the tsunami data assimilation approach and propose a real-time tsunami detection algorithm.

First, I propose GFTDA to increase the assimilation speed. Green's functions between the offshore observational stations and coastal points are calculated and stored in advance. During the assimilation process, tsunami waveforms are forecasted by the superposition of Green's functions without calculating the tsunami wavefield of the entire region. Hence, it greatly reduces the real-time tsunami forecasting time and enables the use of more complicated linear models. The application to the 2012 Haida Gwaii earthquake reveals that GFTDA achieves the same accuracy as the previous data assimilation approach while reducing the time required to issue a valid tsunami warning. The application to the 2004 off the Kii Peninsula earthquake shows that tsunami data assimilation based on the linear DSP model could predict the tsunami arrival time more accurately than the LLW model. It is also successfully applied to the records of the real-time observation network of the 2015 Torishima volcanic tsunami earthquake. The relationship between the accuracy and station number is discussed. Depending on the requirement of forecast accuracy, more attention should be paid to some specific observational stations that play an important role in tsunami data assimilation.

Second, I introduce the concept of virtual stations to conduct tsunami data assimilation for regions with sparse observations. I produce artificial waveforms at virtual stations by interpolating real data of neighboring real stations. I demonstrate the use of this method for the synthetic 2004 Sumatra–Andaman earthquake. The tsunami waveforms at the coasts of India, Sri Lanka, and the Maldives can be forecasted with greater than 70% accuracy. The application of this method to the 2009 Dusky Sound, New Zealand earthquake suggests that the method overcomes the problem of insufficient observations and improves the accuracy from 52.5 to 83.8%. Moreover, it can forecast the far-field tsunamis in the east Pacific generated by the earthquakes in the Chilean subduction zone. The DART tsunameters along the Chilean subduction zone, together with virtual stations, can form a dense network with a substantial spatial coverage. The application of this method to the 2015 Illapel earthquake achieves an

Y. Wang, *Tsunami Data Assimilation for Early Warning*, Springer Theses,
https://doi.org/10.1007/978-981-19-7339-0_7

accuracy of 87.5% in the east Pacific. The adoption of virtual stations can reduce the investment for tsunami disaster mitigation.

Third, I propose a real-time tsunami detection algorithm based on EEMD to detect tsunamis without tidal prediction. In the application of the method to the 2016 Fukushima earthquake (M 7.4), the algorithm can detect a tsunami in real time with a detection delay of less than 1 min (i.e., less than 1/10 of the tsunami wave period). The algorithm could also detect the tsunami signals for the extremely large tsunami from the 2011 Tohoku earthquake (M 9.0) but fails to detect a micro-tsunami from the 1998 off Sanriku earthquake (M 6.4). Whereas the Tohoku earthquake generated a devastating tsunami, the off Sanriku earthquake did not. Hence, the algorithm can greatly improve the efficiency of tsunami early warning system, especially when OBPGs are located close to the coasts of interest. Although there are slight differences between the amplitudes of real-time detected results and the post-processed waveforms, the algorithm can sufficiently determine the tsunami arrival, especially for tsunamis with the potential for substantial inundation. Additionally, the algorithm does not require any source information, and can thus be applied to the tsunami data assimilation approach for providing an early warning for non-earthquake tsunamis.

Finally, I combine the tsunami data assimilation approach with the real-time tsunami detection algorithm. I assimilate the S-net pressure gauge records of the 2016 Fukushima earthquake to forecast the tsunami waveforms at tide gauges in the Sanriku region. To process the raw records for assimilation, I adopt the real-time tsunami detection algorithm based on EEMD to imitate real-time operations. I also compare this method with the traditional post-processing method. The assimilation results of both input time series forecast the tsunami waveforms accurately. The forecast scores of the post-processing method and real-time detection algorithm are 60% and 74%, respectively, for a 35 min assimilation time window; these values improve to 89% and 94%, respectively, on neglecting the stations with imperfect modeling or insufficient offshore observations. Therefore, the combination facilitates a satisfactory tsunami forecast and enables the establishment of a real-time tsunami early warning system. I design a tsunami early warning system based on the data assimilation approach for Japan. It can forecast the tsunami independent of the source, and it can be an important supplement of the traditional tsunami early warning system. Apart from tsunami early warning, in the future, the data assimilation approach can be applied to understand the tsunami propagation processes. The tsunami source characteristics can also be studied through wavefield reconstructing. For example, after an earthquake/landslide occurs, data assimilation may help reconstruct the tsunami wavefield by assimilating offshore data. Then, we can use tsunami inversion method to estimate the tsunami source geometry. It can help us understand the source characteristics (i.e., length, width, slip).

Curriculum Vitae: Yuchen Wang

Education:

Ph.D., Earth and Planetary Science, The University of Tokyo 2021/09.
M.A., Earth and Planetary Science, The University of Tokyo 2018/08.
B.A., Physics (Honored Degree), Peking University 2016/07.

Research Experience:

Japan Agency for Marine-Earth Science and Technology

Postdoctoral Researcher, JAMSTEC Young Research Fellow 2021/10-Present.

Earthquake Research Institute, The University of Tokyo

Graduate Student, JSPS Research Fellow, Kenji Satake's Lab 2016/09-2021/09.

National Research Center for Integrated Disaster Risk Management of Chile

Research Intern, Rodrigo Cienfuegos's Lab 2018/10-11, 2019/10, 2020/03.

Y. Wang, *Tsunami Data Assimilation for Early Warning*, Springer Theses,
https://doi.org/10.1007/978-981-19-7339-0

Laboratory for Climate and Ocean–Atmosphere Studies, Peking University
Research Intern, Yongyun Hu's Lab & Jintai Lin's Lab 2014/04–2016/08.

Honors:

The Director Award of Graduate School of Science 2022/03.
AOGS2021 Best Student Poster Award 2021/08.
2019 SSJ Fall Meeting Outstanding Student Presentation Award 2019/09.
JpGU Meeting 2019 Outstanding Student Presentation Award 2019/05.
JSPS Research Fellowship for Young Scientists 2019/04–2021/09.
GSGC Scholarship 2016/09–2019/03.

Selected Publications:

[1] **Wang, Y.**, Heidarzadeh, M., Satake, K., & Hu, G. (2022). Characteristics of two tsunamis generated by successive Mw 7.4 and Mw 8.1 earthquakes in Kermadec Islands on March 4, 2021. *Natural Hazards and Earth System Sciences*, **22**, 1073–1082. https://doi.org/10.5194/nhess-22-1073-2022.

[2] **Wang, Y.**, Su, H., Ren, Z., & Ma, Y. (2022). Source properties and resonance characteristics of the tsunami generated by the 2021 M 8.2 Alaska earthquake. *Journal of Geophysical Research: Oceans*, **127**, e2021JC018308. https://doi.org/10.1029/2021JC018308.

[3] **Wang, Y.**, & Satake, K. (2021). Real-Time Tsunami Data Assimilation of S-Net Pressure Gauge Records during the 2016 Fukushima Earthquake. *Seismological Research Letters*, **92**, 2145–2155. https://doi.org/10.1785/0220200447.

[4] **Wang, Y.**, Zamora, N., Quiroz, M., Satake, K., & Cienfuegos, R. (2021). Tsunami resonance characterization in Japan due to trans-Pacific sources: Response on the bay and continental shelf. *Journal of Geophysical Research: Oceans*, **126**, e2020JC017037. https://doi.org/10.1029/2020JC017037.

[5] **Wang, Y.**, Heidarzadeh, M., Satake, K., Mulia, I. E., & Yamada, M. (2020). A tsunami early warning system based on offshore bottom pressure gauges and data assimilation for Crete Island in the eastern Mediterranean basin. *Journal of Geophysical Research: Solid Earth*, **125**, e2020JB20293. https://doi.org/10.1029/2020JB020293.

[6] **Wang, Y.**, Satake, K., Maeda, T., Shinohara, M., & Sakai, S. (2020). A Method of Real-Time Tsunami Detection Using Ensemble Empirical Mode Decomposition. *Seismological Research Letters*, **91**, 2851–2861. https://doi.org/10.1785/0220200115.

[7] **Wang, Y.**, Satake, K., Sandabata, O., Maeda, T., & Su, H. (2019). Tsunami data assimilation of cabled ocean bottom pressure records for the 2015 Torishima volcanic tsunami earthquake. *Journal of Geophysical Research: Solid Earth*, **124**, 10,413–10,422. https://doi.org/10.1029/2019JB018056.

[8] **Wang, Y.**, Satake, K., Cienfuegos, R., Quiroz, M., & Navarrete, P. (2019). Far-field tsunami data assimilation for the 2015 Illapel earthquake. *Geophysical Journal International*, **209**, 514–521. https://doi.org/10.1093/gji/ggz309.

[9] **Wang, Y.,** Maeda, T., Satake, K., Heidarzadeh, M., Su, H., Sheehan, A. F., & Gusman, A. R. (2019). Tsunami data assimilation without a dense observation network. *Geophysical Research Letters*, **46**, 2045–2053. https://doi.org/10.1029/2018GL080930.

[10] **Wang, Y.**, Satake, K., Maeda, T., & Gusman, A. R. (2018). Data assimilation with dispersive tsunami model: a test for the Nankai Trough. *Earth, Planets and Space*, **70**:131. https://doi.org/10.1186/s40623-018-0905-6.

[11] **Wang, Y.**, Satake, K., Maeda, T., & Gusman, A. R. (2017). Green's Function-based Tsunami Data Assimilation (GFTDA): a fast data assimilation approach toward tsunami early warning. *Geophysical Research Letters*, **44**, 10282–10289. https://doi.org/10.1002/2017GL075307.

[12] Ren, Z., Hou, J., Wang, P., & **Wang, Y.** (2021). Tsunami resonance and standing waves in Hangzhou Bay. *Physics of Fluids*, **33**, 081702. https://doi.org/10.1063/5.0059383.

[13] Ren, Z., **Wang, Y.**, Wang, P., Hou, J., et al. (2020). Numerical study of the triggering mechanism of the 2018 Anak Krakatau tsunami: eruption of collapsed landslide? *Natural Hazards*, **102**, 1–13. https://doi.org/10.1007/s11069-020-03907-y.

[14] Heidarzadeh, M., **Wang, Y.**, Satake, K., & Mulia, I. E. (2019). Potential deployment of offshore bottom pressure gauges and adoption of data assimilation for tsunami warning system in the western Mediterranean Sea. *Geoscience Letters*, **6**:19. https://doi.org/10.1186/s40562-019-0149-8.

ORCID iD: 0000-0002-0262-1869.

Reviewer for

Nature Communications, Journal of Geophysical Research: Solid Earth, Journal of Geophysical Research: Oceans, Seismological Research Letters, Natural Hazards and Earth System Sciences, Scientific Reports, Natural Hazards, Ocean Dynamics, Geoscience Letters, Pure and Applied Geophysics.

Wolfram [illegible]

The manufacturer's authorised representative in the EU is Springer Nature Customer Service Centre GmbH, Europaplatz 3, 69115 Heidelberg, Germany. If you have any concerns regarding our products, please contact ProductSafety@springernature.com

Printed and bound by CPI Group (UK) Ltd, Croydon, CR0 4YY
30/09/2026
02225259-0002